SEISMIC INSTRUMENTATION DESIGN

Selected Research Papers on Basic Concepts

R. Attri Instrumentation Design Series
(Seismic)

Dr. Raman K. Attri

ISBN: 978-981-11-9751-2 (e-book)
ISBN: 978-981-14-0347-7 (paperback)

First published: 2005
Revised: 2018
Edition: 2nd
Lead author: Raman K. Attri
Published by Speed To Proficiency Research: S2Pro©
Published at Singapore
Printed in the United States of America

National Library Board, Singapore Cataloguing in Publication Data:

Names: Attri, Raman K., 1973-
Title: Seismic instrumentation design : selected research papers on basic concepts / Dr Raman K. Attri.
Description: 2nd edition. | Singapore : Speed To Proficiency Research, [2018] | Series: R. Attri instrumentation design series (Seismic) | Includes bibliographic references.
Identifiers: OCN 1066231381 | | ISBN 978-981-14-0347-7 (paperback) | ISBN 978-981-11-9751-2 (e-book)
Subjects: LCSH: Seismology--Instruments--Design and construction.
Classification: DDC 551.220287--dc23

 Speed To Proficiency
RESEARCH

Speed To Proficiency Research: S2Pro©
A research and consulting forum
Singapore 560463
https://www.speedtoproficiency.com
rkattri@speedtoproficiency.com

To my mother, father, sister and brother who underwent several financial and emotional hardships while supporting my journey to become an engineer and scientist. Humble dedication to such long-lasting hopes of poor parents and siblings for a better tomorrow.

R. Attri Instrumentation Design Series
(Seismic)

CONTENTS

ABOUT THE BOOK

This book is a collection of three papers authored by Dr. Raman K Attri between 1999 to 2001. The book presents early-career scientific work by the author as a scientist at a research organization. The book provides a theoretical and conceptual understanding of concepts and principles for detection and measurements of the seismic signals. The earthquake phenomenon is one of the most unpredictable and often devastating natural events. Sophisticated and advanced technologies are being used for monitoring the seismic activities across the world and efforts are being put in place to develop prediction models. The theory behind the design of sensors, instrumentation and monitoring system is usually not known to electronics and software engineers upfront. The papers included in this book provide such basic guidance to electronics and software design engineers and equip them with the key computational and algorithmic principles based on the underlying theory of seismic activities. These design techniques are fundamental to designing sophisticated seismic instrumentation and earthquake monitoring systems.

The first paper presents a simplified mathematical framework of the seismic events and backend computational software logic that will enable software engineers to develop a customized seismic analysis and computation software.

The second paper presents a simplified description of various earthquake parameters of interest to a seismologist and how these complex parameters are computed using equations.

In the third paper, a visionary concept is presented to integrate geo-scientific instrumentation equipment such as seismic measurement systems to information technology network that would create a centralized web-enabled database that would allow transmitting the data acquired by geographically distributed but networked observatories to better predict or alert about the phenomena like earthquakes.

A NOTE TO READERS

The research papers in this series were authored between 1999 to 2001. As such these papers should be read remembering the time frame in which those were written. Though the basics of design are universal, the author has made no claim regarding the contemporariness of the concepts. While the book presents the most fundamental and universally applicable basic principles in electronics, new advances in electronics and software design should be considered when using or extending the designs discussed in this book.

Each paper was written for different aspects of the overall system design and has been used in this book as-it-is. A substantial overlap, repetitions of text and redundancies are thus imperative to make each paper read of its own.

CITATION DETAILS

The collection can be cited as:

Attri, RK 2018, *Seismic Instrumentation Design: Selected Research Papers on Basic Concepts*, R.Attri Instrumentation Design Series (Seismic), ISBN 978-981-11-9751-2, 2nd edn, Speed To Proficiency Research: S2Pro©, Singapore.

This series contains these seven papers, which can be individually cited as:

Attri, RK 2018/2005, 'Backend Framework and Software Approach to Compute Earthquake Parameters from Signals Recorded by Seismic Instrumentation System,' R. Attri Instrumentation Design Series (Seismic), Paper No. 1, *Seismic Instrumentation Design: Selected Research Papers on Basic Concepts*, ISBN 978-981-11-9751-2, 2nd edn, pp. 1-25, Speed To Proficiency Research: S2Pro©, Singapore.

Attri, RK 2018/2001, 'A simplified Overview: How are the Earthquake Parameters Computed from the Recorded Seismic Signals?', R.Attri Instrumentation Design Series (Seismic), Paper No. 2, *Seismic Instrumentation Design: Selected Research Papers on Basic Concepts*, ISBN 978-981-11-9751-2, 2nd edn, pp. 27-52, Speed To Proficiency Research: S2Pro©, Singapore.

Attri, RK 2018/1999, 'GSIS: A Conceptual Model for Web-based Integration of Information Technology with Geoscientific Instrumentation,' R.Attri Instrumentation Design Series (Seismic), Paper No. 3, *Seismic Instrumentation Design: Selected Research Papers on Basic Concepts*, ISBN 978-981-11-9751-2, 2nd edn, pp. 53-75, Speed To Proficiency Research: S2Pro©, Singapore.

Author's previous research work on snow hydrology can be cited as:

Kumar, S, Attri, RK, Sharma, BK & Shamshi, MA, 2000, 'Software Tool for

Seismic Data Recorder and Analyser', *IETE Journal of Education*, vol. 41, No. 1-2, pp. 23-30, https://doi.org/10.1080/09747338.2000.11415718, available at https://www.tandfonline.com/doi/abs/10.1080/09747338.2000.11415718, https://www.researchgate.net/publication/277592177

ABBREVIATIONS

EMF	Electro-magnetic force
FTP	File Transfer Protocol
GSIS	Geo-Scientific Information System
GUI	Graphical user interface
HTML	Hyper-Text Markup Language
IRIS	Incorporated Research Institutions for Seismology
ISP	Internet Service Provider
IT	Information Technology
LTA	Long-term average
LAN	Local Area Network
MSDOS	Microsoft Disk Operating System
MDQP	Multi-dimensional query Processor
PEM	Pre-event minutes
PET	Post-event time
PPP	Point Protocol
RTMS	Real-Time monitoring system
SLIP	Serial Line Internet Protocol
STA	Short-term average
RDBMS	Relational Database management System
WAN	Wide Area Network

BACKEND FRAMEWORK AND SOFTWARE APPROACH TO COMPUTE EARTHQUAKE PARAMETERS FROM SIGNALS RECORDED BY SEISMIC INSTRUMENTATION SYSTEM

RAMAN K. ATTRI

Ex-Scientist (Geo-Scientific Instrumentation)
Central Scientific Instruments Organization INDIA

Abstract: Digital data acquisition systems empower computation of seismic parameters and its interpretation from the recorded earthquake signal. This enables a seismologist to compute all the relevant parameters automatically. Futuristic applications require extensive software development to implement seismic prediction and forecasting models. While developing such models, software developer prefers to write their own in-house analysis & modeling software with complete control over the required computations and models. This paper presents a simplified mathematical framework of the seismic events and backend computational software logic & algorithm to provide a simple framework for software engineers develop customized seismic analysis & computation software.

Index Terms: Earthquake signals, Seismic Instrumentation, Earthquake monitoring, Software approach to seismic measurements, seismic parameters

I.   INTRODUCTION

THE seismological study is closely linked with implementing right kind of the seismic instrumentation for recording these earthquakes, interpreting them, storing their history over the years. Software systems enhance the power of such instrumentation by performing the major job of signal analysis, complex computation of parameters, data interpretation, fetching inference, statistical trend analysis of seismic activity at a place of interest over the years. In association with instrumentation systems, the complex software system deployment sometime ensures right forecasting and generating warning systems for an earthquake. This makes the job of a seismologist more accurate, objective, automated and quick.

Although a range of off-the-shelf software is available in the market, however, those sometimes fail to address the needs of futuristic modeling. For a seismologist to develop a prediction model, they need software to compute the basic seismic parameters and a framework upon which they can develop their prediction or forecasting model [6]. Software developers typically are not seismologists and are not aware of the complex mathematics behind the geophysical activities. This paper describes the simplified backend software approach to enable such developers to develop a software framework to compute and interpret seismic parameters. This paper outlines basic logic/ algorithmic approaches build on mathematically equations integrated into software algorithms to correlate data points in seismic signal together to

II.   SEISMIC RECORDING & ANALYSIS CHAIN

When an earthquake occurs, it generates an expanding wavefront from the earthquake hypocenter at a speed of several kilometers per second [22]. Generally, it takes few second for theses waves front to travel thousands of kilometers [20]. A wavefront expansion is shown in Fig. 1. This wavefront consists of two unique waves: One P-wave which comes earlier due to its faster speed and another S-wave front which comes later due to the little slower speed of travel [16]. The P-wave front is released first by the earthquake reach the seismic station (shown as 'A') and S-wave front soon follow the P-wave front.

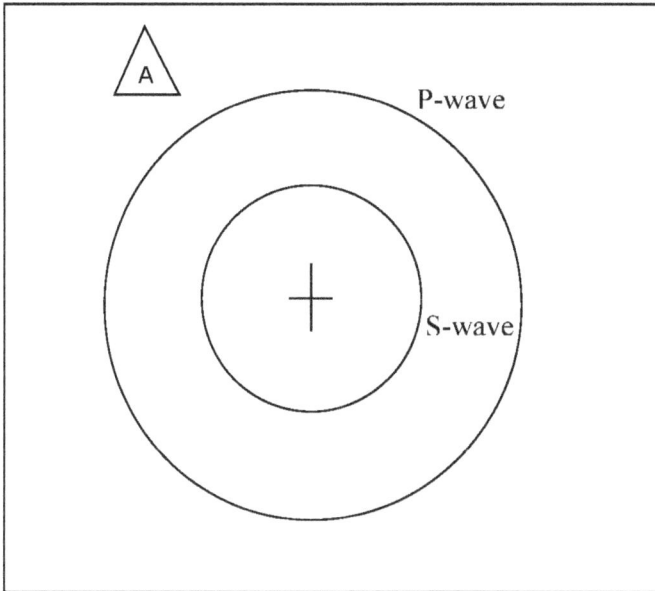

Fig. 1. The expanding circles indicate the expanding wavefronts at every subsequent second after the earthquake is originated at hypocenter. S-wave front following the P-wave front in an actual earthquake

Thus, we get two unique wave peaks on the recording instruments, each corresponding to its wavefront. A typical waveform recorded on a true earthquake helped by a seismic sensor is shown in Fig. 2 [6].

Fig. 2. A typical seismic signal recorded during an earthquake has two distinct peaks spaced with an interval. The first peak corresponds to P-wave and the second peak corresponds to S-wave. The S-wave is the seismic after-shock which causes more impact and consists of peaks.

It can be seen that the signal contains a lot of residual background noise and high peaks of the earthquake event. Normally an amplifier with a wide dynamic gain/range (> 120 dB) must faithfully amplify the signal from the sensor [13].

Earthquakes are monitored with a network of seismometers on the earth's surface. Since no single instrument can operate over a wide bandwidth and dynamic range, therefore, a set of instruments for different bandwidth are needed to operate in conjunction [12]. The ground motion at each seismometer is amplified and recorded electronically at a central recording site. As the wavefront expands from the earthquake, it reaches more distant seismic stations [22]. Seismic data obtained from many stations must be correlated. The data in digital format is downloaded from the system to a PC helped by interfacing software. This raw data plays an important role in further seismic analysis, interpretation and prediction modeling.

III. SOFTWARE APPROACH TO SEISMIC ANALYSIS

Seismic data acquisition systems usually come with the interface, downloading and analysis software. If not, the research engineer needs to develop their own data acquisition and seismic analysis software. Seismic Data Analysis and Interpretation Process begins with raw data retrieved from the data acquisition system and downloaded into a PC. Specialized software is used to process the data, retrieve parametric information and frame inferences.

Few characteristics are of the interest to a seismologist [14]. A seismologist would be interested in the following parameters:

Timing Parameters:
    The arrival time of P and S-wave (Time of occurrence)
    Coda length (total event duration)
Location Parameters:
    Focal Point (where the earthquake originated)
    Epicentral distance (point above focal point)
    Focal length (depth of origin)

Magnitude Parameters:
    Ritchet Scale,
    Coda Magnitude
Intensity Parameters:
    Intensity/ Energy
    Moment
    Ground motion

This computation is done by seismic processing software. As shown in Fig. 3, the input to the software is the raw analog seismic signal digitized by signal digitizer in real-time. Event detection module selectively detects the seismic events and the framework computes time, location, magnitude and energy parameters of the seismic event. The framework also contains a display and analysis module. Advanced software framework interfaces the output to prediction and forecast modeling software.

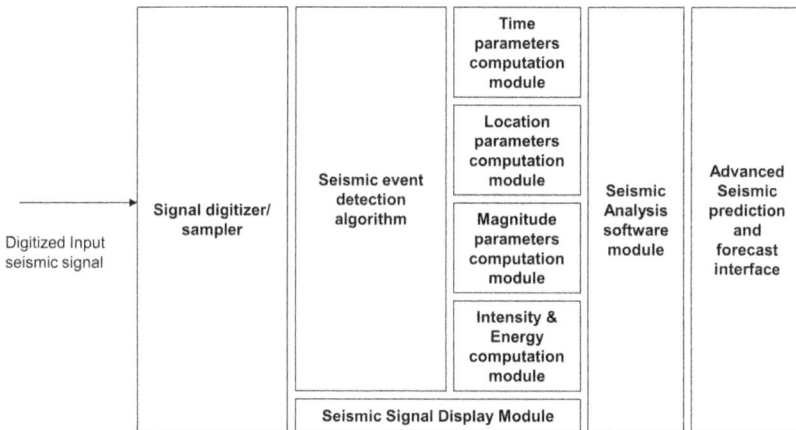

Fig. 3. Software-based processing and analysis framework for seismic parameters computation and interpretation.

IV. COMPUTING TIMING PARAMETERS USING SOFTWARE LOGIC

The timing parameters in seismology are taken regarding the arrival of P- and S-wave. Generally, there are two types of timing parameters namely: S-P time interval and Coda Length which are of prime importance to seismologists [7]. These are depicted in Fig. 4.

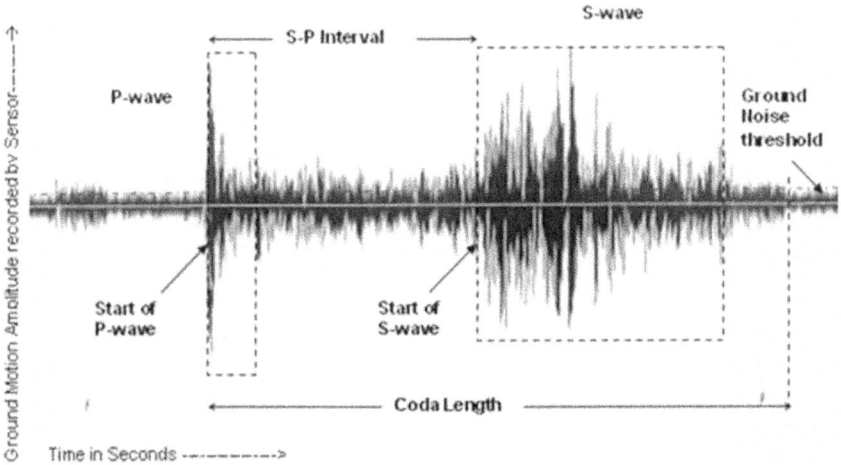

Fig. 4. Time Parameters of a seismic signal on a recorded seismogram. Difference between the arrival time of P-wave and S-wave is SP-interval and time from arrival of P-wave till the settlement of ground motion in an after-shock is called coda length.

A. Identifying Arrival Time for P and S-Waves Using Software Logic

In software, peak amplitude capturing algorithms is used to detect the arrival of first P-wave. However, before capturing the arrival time of P- and S-wave, it is crucial to detect the occurrence of the earthquake.

1. STA- LTA Ratio Software Algorithm to Detect Seismic Event Occurrence: One of the software techniques to detect earthquake selectively and to isolate it from extraneous events is STA-LTA averaging. The method involves the computing of long-term and short-term averages. In this algorithm, samples of data are stored well before the current time in history, and an average is computed over a long period. This is the

average ground noise level persisting at that location at all time when there is no seismic activity. This is termed as long-term average (LTA). For calculating LTA, data samples of the past few seconds (say Y seconds) are accumulated. The value Y varies from 1 to a few tens of seconds [18].

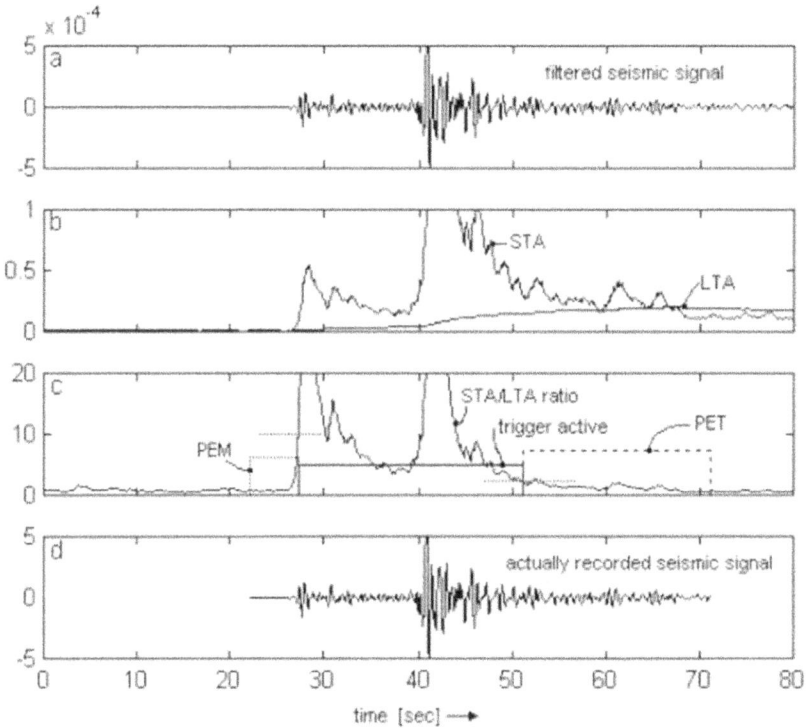

Fig. 5. The concept of LTA/ STA computation on seismic signal. A) Filtered seismic signal b) STA and LTA averages calculated on input signal c) Trigger activated based on a set STA/ LTA trigger threshold. Data acquisition system records the seismic event data including PEM (pre-event minutes) before the trigger until PET (post-event time) after the trigger d) actual recorded signal based on PEM and PET values (Courtesy: Trnkoczy, 1998, Kinematrics Inc [18])

The sample average is also calculated in small time frames. These small time frames are usually selectable based on the seismic profile of the location. Usually, these are the short duration in the range of a few one-tenth of a second (say 0.X seconds). This is termed as short-term average (STA). Modern data acquisition systems have circular memory implemented in it to record the historical data samples prior to current time samples.

The moment the earthquake is occurring the short-term average is expected to be more than the long-term average. A suitable threshold point is chosen depending upon general ground motion history of the site (termed as an alpha ratio) as shown in figure 5. The moment the ratio of STA and LTA cross this threshold, the earthquake is supposed to have occurred. Generally, a ratio of > 1 is a good indicator of earthquake activity and once triggered the system keeps recording the earthquake event until the end of the activity. This approach ensures that we have sufficient data before and after the earthquake to carry pre-earthquake studies too. Taking the average is to rule out some sudden spikes in the signals due to some artificial means or some other ways. This approach will ensure that only seismic data is recorded and hence P and S-wave is detected for a true seismic event only [1].

2. Peak Amplitude Vs Ground Noise Algorithm: Another technique to detect P and S-peaks is to detect the peaks only with a peak detector and finding the ratio of the peak amplitude with the average noise level. The peak capturing generally indicate the arrival of P-wave and S-wave. In this algorithm, software is configured to spot the beginning of such P-waves or S-waves through discrimination and adaptive differential of adjacent samples to know when a particular wave has started. These algorithms selectively detect the arrival of P-wave and S-wave.

Both of the above algorithms can be implemented simultaneously in the software. This has the advantage that software accurately pinpoints the arrival time of P-wave and S-wave separately and the computation will be for true earthquake signal only.

B. Computing S-P Time Interval Using Software Logic

This is the difference between the arrival time of P and arrival time of S. The seismologists analyze the S-P time difference data over the years for a particular location and can come up with minimum time difference available between reception of P-wave and actual heavy disastrous S-wave. How closely S and P-waves follow each other provides important inferences regarding the creation of some early warning systems for some strategic locations.

Once the time of arrival of P-wave and S-wave are marked, the S-P time interval can be computed accurately. For this, the clocks of most the seismic data acquisitions systems are synchronized with some universal timing standards like WWV or ATA.

S-P interval (in sec) = Absolute arrival time of S-wave
– Absolute arrival time of P-wave
- eq (1)

C. Computation of Coda Length Duration Using Software Logic

Coda length is basically the time in seconds starting from P-wave time to the end of a wave up to noise level. This time indicates how long a seismic shock prevailed and how much time was taken by the earth to calm down after the shock. This time is critical in relating the amount of destruction with the time interval of the earthquake. This time measurement starts from right from the occurrence of the P-wave. For a given earthquake, coda length is measured by different stations and the average value of these durations is taken [17].

Coda Length duration is computed by software after ascertaining the start of the P-wave till the waveform amplitude and pattern match with what it was before the start of P-wave. From measurement and electronics system perspective, all-time prevailing ground motion converted into induced EMF in the sensor can be taken as the measure of prevailing noise. This noise can be further averaged over the time. A software system computes the average value of noise samples to specify 'settled' ground activities prevailing at all time. Refer to Fig. 4 to see Coda length marked on the seismogram recording.

The coda length can be found out by clubbing the algorithm which detects the P-wave arrival time with the algorithm which detects the prevailing noise level. When the signal amplitude and earth vibration falls below the stipulated threshold, the earthquake is said to be settled and that time is found out by subtracting the time of arrival of S-wave from the time of settlement of an earthquake.

Coda length (in seconds) = Absolute time of settlement
– Absolute time of arrival of P-wave
- eq (2)

V.   COMPUTING LOCATION PARAMETERS USING SOFTWARE LOGIC

Two important parameters in locating the earthquake are: Epicenter and hypocenter. The epicenter is the point exactly above the earthquake on the earth surface. The point where the earthquake generates is called hypocenter. Hypocenter is also called the focal point sometimes; this is shown in Fig. 6. Epicentral Distance is the straight-line distance between the point of observation of an earthquake and the point exactly above the focal point on the earth surface. This is measured in degrees and can be converted in kilometers depending upon the parameter constants of the location where the earthquake is being measured [17].

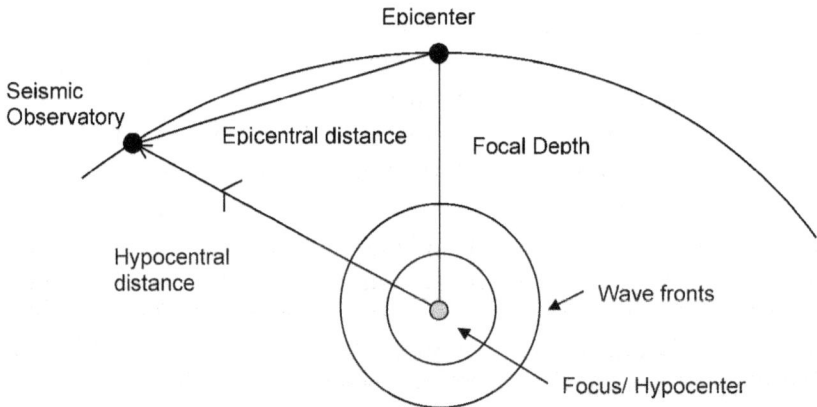

Fig. 6. The point where an earthquake is originated is called focus or hypocenter. The point directly above this on the earth surface is called Epicenter.

A.   Using S-P Time Interval to Find Epicentral Distance (Location of Earthquake) Through Software Logic

1. Backend Mathematical Framework: the earthquake is measured by a network of seismic stations located at different locations. When an earthquake occurs, we observe the times at which the wavefront passes each station [22]. We find the unknown earthquake source by knowing these wave arrival times at different seismic stations [10]. Since P-waves travel faster than S-waves, the time difference between the arrival of the P-wave and the arrival of the S-wave depends on the distance the waves traveled from the source (earthquake) to the station (seismograph).

Louie [10] and IRIS white paper [19] present a method of triangulation to compute earthquake location. This is an excellent method which is used in the software logic. Software logic uses the fact that P and S-waves each travel at different speeds and therefore arrive at a seismic station at different times and time interval between arrival of S and P-wave can be used to find the location of the earthquake [19]. P-waves travel the fastest, so they arrive first. S-waves, which travel at about half the speed of P-waves, arrive later. A seismic station closes to the earthquake records P-waves and S-waves in quick succession. With increasing distance from the earthquake, the time difference between the arrival of the P-waves and the arrival of the S-waves increases [19]. This concept is made more apparent in Fig. 7, where three different seismic stations located at a different distance record the same event.

IRIS white paper [19] outlines a method of triangulation to computer epicentral distance/ location from recorded seismogram. The IRIS white paper illustrates the earthquake occurred in Mexico (Courtesy: IRIS Consortium) to illustrate the computation of epicentral distance [19]. The original signal images have been reproduced here in Fig. 8. The arrival time of P-wave, and S-wave, as well as S-P interval, is different at three locations depending upon the distance from the earthquake.

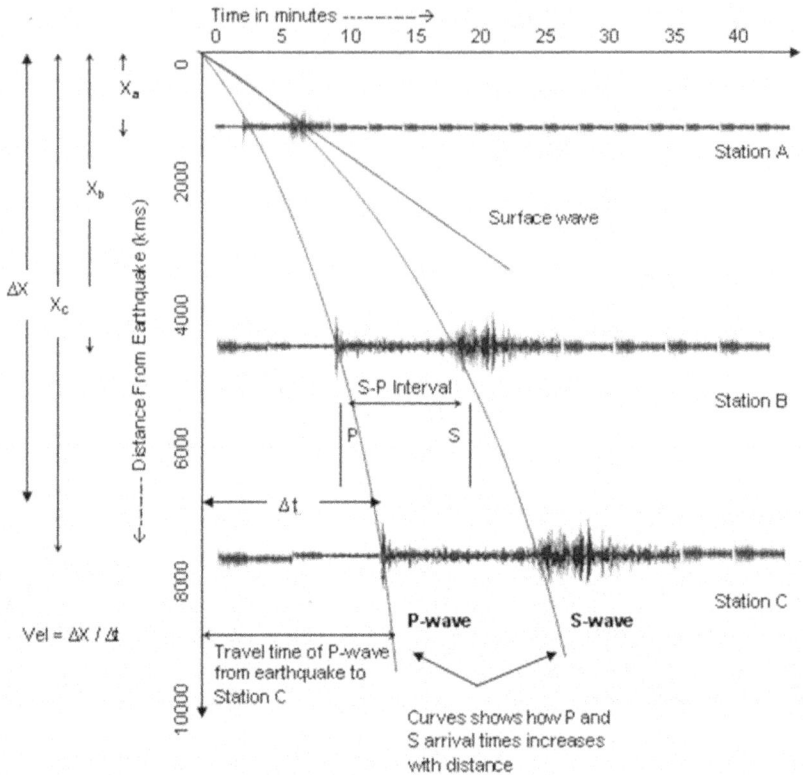

Fig. 7. The same earthquake recorded by three distant stations where P-wave and S-wave arrive at different times. The velocity of P-wave and S-wave can be computed from these three data points. S-P interval is used to calculate the distance of the station from the point of earthquake origin [10, 20].

From observing and analyzing many earthquakes, we know the relationship between the S-P time and the distance between the station and the earthquake. We can, therefore, convert each measured S-P time to distance [19]. In software, a simple lookup table can be made to simplify such computations. A time interval of 1.5 minutes corresponds to a distance of 900 kilometers, 3 minutes to 1800 kilometers, and 5 minutes to 3300 kilometers [19]. Multiply the seconds of S-P time by 8 km/s for the kilometers of distance [10].

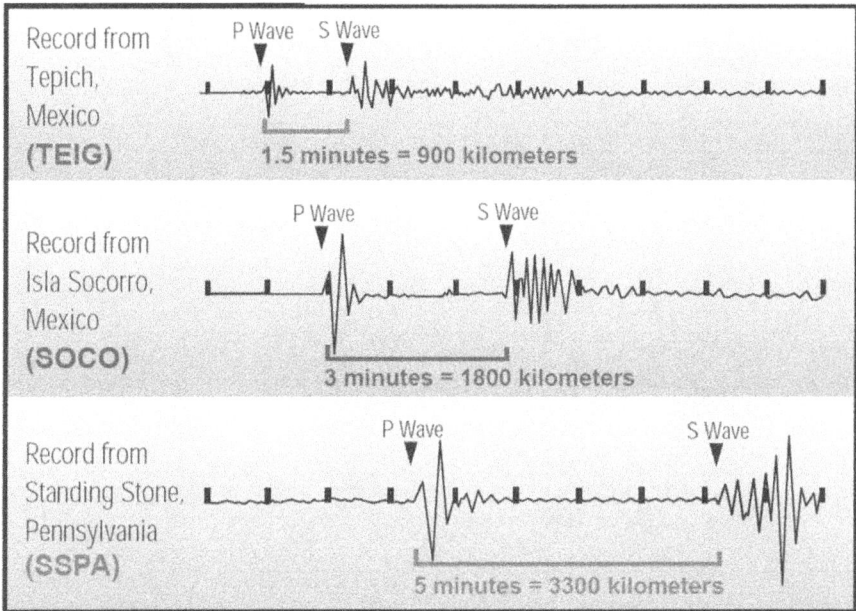

Fig. 8. The seismic signal recorded at three stations from the same earthquake. SP-interval increases as the distance of the station increase from the earthquake epicenter. (Image reproduced with permission from IRIS [19])

2. Software Implementation: Once the distance to the earthquake for three stations is known, the location of the earthquake can be determined using software logic. If an earthquake occurred a given distance from a station, it could have been anywhere on a circle whose radius is that distance, centered on the station. If distances from two stations are known, two locations are possible: the two intersection points of two circles. If distances from three stations are known, the earthquake can be unambiguously located. This is the principle of triangulation. For each station, we draw a circle around the station with a radius equal to its distance from the earthquake [19].

The earthquake occurred at the point where all three circles intersect as found in Fig. 9. In software, this method of locating the earthquake from data available from at least three sources can be programmed in an algorithm using built-in functions in the programming.

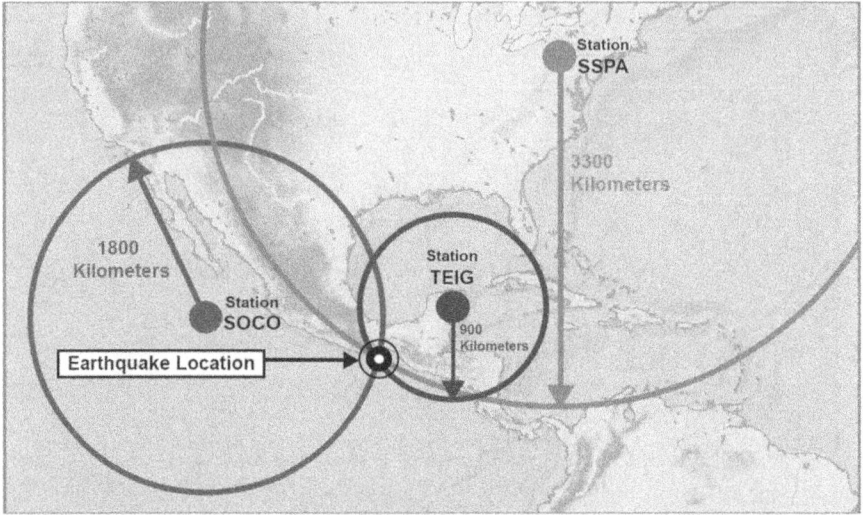

Fig. 9. Method of Circle intersection to find earthquake location. A circle is drawn around distance translated from SP-interval, and the intersection of the circle is an actual location or epicenter of the earthquake (Images reproduced with permission from IRIS [19])

The algorithm is simple to state: guess a location, depth and origin time; through the software itself compare the predicted arrival times of the wave from your guessed location with the observed times at each station; then move the location a little in the direction that reduces the difference between the observed and calculated times [22]. Then this procedure is repeated by the software, each time getting closer to the actual earthquake location and fitting the observed times a little better [22]. Software stops when its adjustments have become small enough and when the fit to the observed wave arrival times is close enough [10]. Above method of triangulation gives the epicentral distance in kilometers.

B. Computing Epicentral Distance Using Lookup Tables in Software Logic

Epicenter Distance in degrees is computed from this S-P time interval. The exhaustive table is available which lists epicentral distance in degrees corresponding to each SP-interval in seconds [3, 4]. Seismologists have generated this lookup table by years of research. The basic concept of these tables is based on the time taken by S-wave to reach after P-wave is directly proportional to the velocity of the ground waves and the distance of the originating point to the point where S-P interval is being observed. This readily available table can be integrated into the software using any database technique in the form of the tabulation of S-interval vs epicentral distance in degrees. Once integrated into the software, the epicentral distance can be determined by the software from the database based on the previously determined S-P interval.

The epicenter distance in kilometers can be found from the epicentral distance in degrees by another lookup table [3, 4]. This table lists the conversion factor that must convert the epicentral distance found in degrees into kilometers. Every location on earth has particular characteristics which defined this constant. For example, for a placed named, Chandigarh (India Latitude: 30° 42' N, Longitude: 76° 54' E) this constant has already been computed by seismologists to be 111.1 as shown in equation (3). This straightforwardly calculates epicentral distance in kilometers once epicentral distance in degrees is known.

$$Epicentral\ Distance\ (in\ Km)\ =\ 111.1\ x\ \Delta \qquad \text{- eq (3)}$$

Where Δ the epicentral distance in degrees calculated using the triangulation method.

This table can also be integrated into the software using database techniques as described earlier. However, the location of the recording is to be stored with the data during data acquisition or it is to be input by the seismologist. Depending upon city string in the input, the software will automatically calculate Epicentral distance in kilometers from computed Epicentral distance in degrees.

Another easy and straightforward technique to find the epicentral distance in kilometers is from equation (4) for which we need to know the velocities of P-wave and S-wave, which are very much known. The equation (4) can find the epicentral distance using S-P time interval. This will save the algorithms and processing needed for lookup table management in software. Normally we know the velocity of P & S-waves. We can calculate the Epicentral distance Δ in kms from the S-P interval as under:

$$\textit{Epicentral Distance in Kms, } \Delta = V_p / (V_p/V_s\text{-}1) (T_s\text{-} T_p) \quad \text{- eq (4)}$$

Where V_p and V_s are the velocities of P-wave and S-wave respectively. T_p and T_s are their respective arrival times, as recorded by the software system.

For crustal rocks, typical V_p is 6 km/s and V_p/V_s is approximately 1.8 km. Epicentral distance in kilometers is about 7.5 times S-P interval in seconds. If three or more Epicentral distances measure by various seismic stations is available, it is possible to use a map to plot these data. The epicenter may be placed at the intersection of circles with stations as the center and appropriate Δ as radii. This point of intersection may be an aerial distance giving the uncertainty in locating exact epicenter.

C. Computing Focal Depth Using Software Logic

Focal Depth is the vertical distance from the focal point to the epicenter (i.e. point exactly above it vertically on the earth surface) is called Focal Depth. This gives the depth of the focus or hypocenter beneath the earth's surface. A distance "Hypocenter Distance" is defined to indicate the distance between focus (or hypocenter) and the point of observation or the point where an earthquake has produced its effect. This is a straight-line distance [17].

At software logic, a triangulation method can compute hypocenter distance and depth. For more sophisticated systems, at the backend, mathematically, the problem is solved by setting up a system of linear equations, one for each station. The equations express the difference between the observed arrival times and those calculated from the previous (or initial) hypocenter in terms of small steps in the 3 hypocentral coordinates and the origin time [24]. We must also have a mathematical model of the crustal velocities (in kilometers per second) under the seismic network to calculate the travel times of waves from an earthquake at a given depth to a station at a given distance. The system of linear equations is solved by the method of least squares which minimizes the sum of the squares of the differences between the observed and calculated arrival times [22]. The process begins with an initially guessed hypocenter, performs several hypocentral adjustments each found by a least squares solution to the equations, and iterates to a hypocenter that best fits the observed set of wave arrival times at the stations of the seismic network [8, 20, 24].

Above software system structure is shown in Fig. 10 which computes timing information, epicentral distance in degrees and kilometers and performs triangulation method to locate the earthquake.

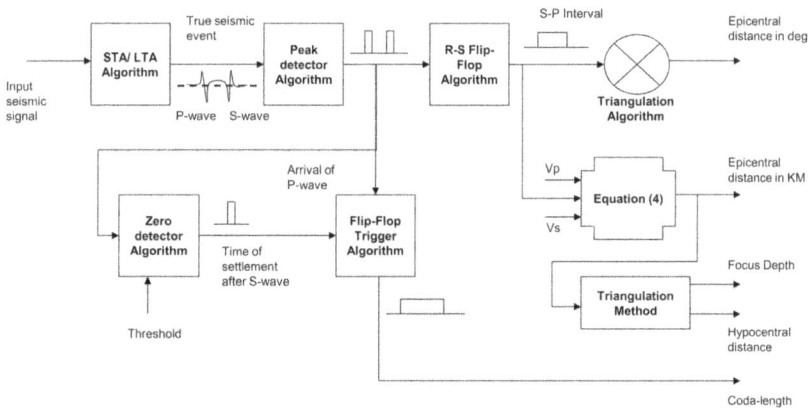

Fig. 10. A comprehensive software framework for locating earthquake

VI. COMPUTING MAGNITUDE PARAMETERS USING SOFTWARE

Although each earthquake has a unique magnitude, its effects will vary significantly according to distance, ground conditions, construction standards, and other factors. A relatively small magnitude earthquake that happens near the surface can cause shaking of great intensity, whereas a large magnitude earthquake that happens in a depth of several hundred kilometers will not necessarily produce intense shaking at the surface. The magnitude on the Richter scale relates to the energy released by the earthquake, and it is independent of possible damages on the surface of the earth.

A. Computing Ritchet Scale Magnitude Using Software Logic

1. Backend Mathematical Framework: Ritchet scale is normally determined based on maximum amplitude on Wood-Anderson seismograph (which have nearly constant displacement amplification over the frequency range of local earthquake). Ritchet scale defined the Local Earthquake Magnitude, observed at the place of observation, denoted by M_L irrespective of the direction and its origin and empirically given by equation (5) as under:

$$M_L \quad = \quad Log\ A - Log\ A_0\ (\Delta) \qquad \text{- eq (5)}$$

Where A, Maximum amplified in millimeters on Wood-Anderson seismographs, Δ, the Epicentral distance in kilometers and A_0 (Δ) is maximum amplitude at Δ kilometers for a standard earthquake.

Local Magnitude is thus a number characteristic of earthquake and independent of the location of the recording seismographs. The second factor was scaled by certain assumptions by Ritchet. To solve the above equation (5), a table of $-log\ A_0$ as a function of Epicentral distance in kilometers is needed. Ratchet arbitrarily chose $-log\ A_0 = 3$ at $\Delta = 100$ Kilometers, and other entries in the table were constructed from observed amplitudes of a series of well-located earthquakes [3, 4].

Gutenberg & Ritchet (1940) extended local scale magnitude of distant earthquakes to call it as Surface Wave Magnitude, denoted by M_S and empirically given in equation (6) as under:

$$M_S \quad = \quad Log\ A - log\ A_0\ (\Delta_0) \qquad \text{- eq (6)}$$

Where Δ_0 is epicentral distance taken in degrees to consider and origin of the origin of the earthquake, A is maximum combined horizontal ground amplitude in micrometers for surface waves with a period of 20 sec and $-log\ A_0$ is tabulated as a function of epicentral distance Δ in degrees, similar to that for local magnitude.

This applied to shallow earthquakes generating observable surface waves. However, for distant earthquakes at various depths inside the earth, only body waves are observable as P and S-waves. So, the magnitude had to be defined based on observed P-wave and S-wave amplitudes. A new relationship for body waves was defined as Body Wave Magnitude, denoted as M_B and empirically given by equation (7) as under:

$$M_B \quad = \quad log\ (A/T) - f\ (\Delta.\ h) \qquad\qquad - eq\ (7)$$

Here (A / T) is the maximum amplitude to period ratio in micrometer per second, F $(\Delta,\ h)$ is a calibrated function of Δ and focal depth h. This function has been characterized for most major locations across the globe and empirically it can be written in equation (8) as:

$$M_B \quad = \quad log\ (A/T) + \Delta + C \qquad\qquad - eq\ (8)$$

Where A is the amplitude of the highest S-wave Peak, T is the time period of the same peak, as shown in Fig. 11. C is place specific constant. This *distance factor* comes from a table that can be found in Richter's [15] book *Elementary Seismology*. Theoretically, it is the logarithmic value of the amplitude of the highest peak with the time period of that peak.

Higher the amplitude, higher the strength of the earthquake and higher the time period lower will be the strength. The sharpest most peaks indicate the highest strength of the earthquake occurrence.

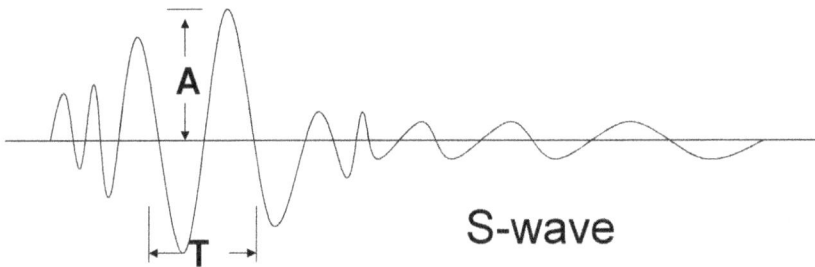

Fig. 11. A (amplitude) and T (time period of highest peak) of S-wave forms the basis of Ritchet scale calculation

2. Software Implementation: In practice, the software will compute the magnitude of Body waves using Ritchet Scale. This magnitude M_B depends upon the maximum amplitude of the signal detected, its frequency and locational characteristics of the site of observation. The Ritchet scale of a recorded earthquake is computed using the equation (8). The location constant is again integrated into the software using the readily available tables. The maximum amplitude 'A' can be determined by the peak detector software algorithm. Computing the time period of the highest peak requires a good amount of processing. This needs detection of zero crossing immediately preceding and succeeding the highest peak detected by the peak detector algorithm. This can be done by incremental time steps below and above the time mark of the peak and marking the time marks where the signal becomes zero on both sides. The difference between these two points is the time period the peak. The Epicentral distance from previous algorithms and constant 'C" value determined from lookup table along with the computed time period enable software to compute the Ritchet scale [3, 4]. Software system architecture to compute amplitude of P-wave, S-wave and local magnitude is shown in figure 12.

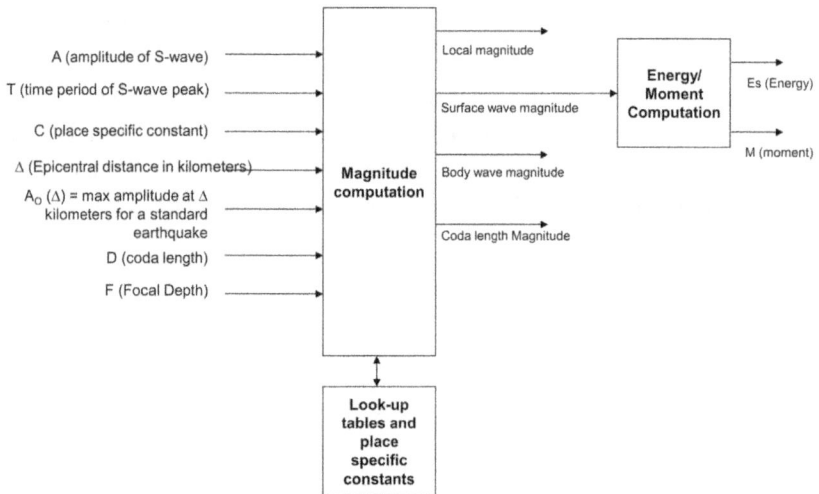

Fig. 12. A software framework for calculating magnitude and energy of the earthquake

B. Computation of Coda Length Magnitude

1. Backend Mathematical Framework: It is an estimate of local magnitude M_L calculated using the coda-length/magnitude relationship [17]. Coda Magnitude is directly related to the coda length and epicentral distance. This approach is based on using the signal duration rather than the maximum amplitude to estimate the earthquake magnitude because it will make it independent of using Wood-Anderson Seismograph. Bisztricsany [12] and Lee et al. [9] correlated Ritchet magnitude with a signal duration of the local earthquake through an empirical formula given in equation (9).

$$M_D \quad = \quad -0.87 + 2 \log D + 0.0035 \, \Delta \qquad \text{- eq (9)}$$

Where   D is Coda length, Δ is epicentral Distance

Similar empirical relationships in many forms integrating many dependent factors have been suggested by many researchers, the more general form of the same is shown in equation (10) as under:

$$M_D \quad = \quad a_1 + a_2 \log D + a_3 \, \Delta + a_4 \, h \qquad \text{- eq (10)}$$

Where h is the focal depth and a_1, a_2, a_3 are empirical constants.

The above equation can be empirically established for a location of interest depending upon certain empirical relationship. For example, the generalized empirical relationship reduces to that given in equation (11) for a city like Chandigarh (India Latitude: 30° 42' N, Longitude: 76° 54' E) as under:

$$M_D \quad = \quad -0.57 + 1.38 \log D + 0.29 \, \Delta \qquad \text{- eq (11)}$$

The underlying concept is that coda magnitude is a measure of the impact of the earthquake which occurs at a specified epicenter in a particular direction, specified epicentral distance and sustains for a specified duration in seconds.

2. *Software Implementation:* Software can easily compute the Coda magnitude from Coda length and Epicentral distance in degrees from equation (10) with appropriate empirical coefficients depending upon the location. These empirical coefficients may be entered by the seismologist or lookup table based on location entered could be provided. Or, the software could work specifically with customized empirical relationship e.g for Chandigarh using equation (11) or generalized equation (9). This coda length is one indicator of the earthquake intensity. The software system to compute coda magnitude is shown in Fig. 12.

VII. COMPUTATION OF SEISMIC ENERGY USING SOFTWARE

A. *Backend Mathematical Framework*

Seismologists use a Ritchet Magnitude scale to express the seismic energy released by each earthquake. Both the magnitude and the seismic moment are related to the energy that is radiated by an earthquake. Richter and Gutenberg (1940) developed a relationship between magnitude and energy. Their relationship is given by equation (12):

$$Log\ ES = 11.8 + 1.5MS \qquad \text{-eq (12)}$$

Where E_S is energy in ergs from the surface wave magnitude M_S. Note that E_S is not the total intrinsic energy of the earthquake, transferred from sources such as gravitational energy or to sinks such as heat energy. It is only the amount radiated from the earthquake as seismic waves, which ought to be a small fraction of the total energy transferred during the earthquake process [11]. The Richter scale is based on the maximum amplitude of certain seismic waves, and seismologists estimate that each unit of the Richter scale is 31 times increase of energy [11].

Seismologists have developed a standard magnitude scale completely independent of the type of instrument. It is called the moment magnitude, and it comes from the seismic moment. Earthquakes are caused by internal torques, from the interactions of different blocks of the earth on opposite sides of faults [11]. The moment of an earthquake is fundamental to seismologists' understanding of how dangerous faults of a certain size is expressed by equation (13) below:

$$M_o \quad = \quad \mu \, A \, d \qquad \qquad - eq \, (13)$$

Where M_o is moment of an earthquake (given in units of dyne-cm), μ is the rigidity of the rocks; A is fault area, D is the slip distance.

Kanamori [5] gave a relationship between seismic moment and seismic wave energy, given by equation (14) below:

$$E_S \quad = \quad M_o / \, 20000 \qquad \qquad - eq \, (14)$$

B. Software Implementation

The seismic energy can be computed by the software using the equation (12). This takes the Surface wave Ritchet magnitude, M_S as input. The surface wave magnitude can be computed by equation (3) where the factor $-\log A_0$ is available in form of exhaustive tables which can be duly entered in software using database approaches. The seismic energy is computed in ergs.

The moment can be calculated from equation (13) if physical characteristics of the earth layer where earthquake erupted are known. Equation (14) is a good approximation for the same. A software system which computes moment and energy from surface wave amplitude shown in Fig. 12.

VIII. CONCLUSION

Software-based approach to seismic analysis exploits the maximum processing capability of the computer. Rather than manually solving complex equations, the function is handled by software. The above-highlighted software approach would have advantages like manual and automatic Data viewing with auxiliary information, correlation of the conclusions and inferences drawn from individual recorded data and identification of the seismicity of the site. This paper is an important research aid for developing customized software to convert manual framework into an automated system which extends its capability to the prediction process.

ACKNOWLEDGMENTS

The author acknowledges the previous work, consulting and feedback provided by following individuals which helped to shape the structure of this paper.

1. Naresh Kumar, Ex-Seismologist, CSIO Chandigarh
2. Satish Kumar, Senior Geo-Scientific Instruments Scientist, CSIO Chandigarh
3. B.K. Sharma, Deputy Director & Senior Scientist, Geo-Scientific Instruments Division, CSIO Chandigarh
4. Incorporated Research Institutions for Seismology (IRIS) Consortium, Washington DC USA. http://www.iris.edu for permissions to use white paper
5. Kinemetrics Inc. http://www.kinemetrics.com for permission to use its white paper

IX. REFERENCES

[1]. Ambuder, B.P. & Soloman, S.C. (1974), An event recording system for monitoring small earthquakes, Bulletin of Seismological Society of America, Vol 64, pp 1181-1188.

[2]. Bisztricsany E. (1958), A new method for the determination of the magnitude of earthquakes, Geofiz. Kozl (Hungary). Vol 7, pp. 69-96.

[3]. Geotechnical Corp (1960), Seismogram Analysis Training Document, Geotechnical Corp, Texas USA, URL: http://psn.quake.net/info/analysis.pdf.

[4]. Harold, J., & Bullen, K.E. (1948), Seismological Tables, Office of the British Association, W.I. London, UK.

[5]. Kanamori, H. (1977), The energy release in great earthquakes: Journal of Geophysics Research., Vol 82, p. 2981-2876

[6]. Kumar, S., Attri, R.K., Sharma, B.K., Shamshi, M.A. (2000), Software Tools for Seismic Signal Analysis, IETE Journal of Education, Vol 41, No. 1& 2, pp 23-30. URL: https://doi.org/10.1080/09747338.2000.11415718.

[7]. Lee, W. H. K. (1987). Observational seismology: in 'Encyclopedia of Physical Science and Technology', Vol 12, pp. 491-518. Academic Press, USA.

[8]. Lee, W. H. K., and Lahr, J. C. (1972b), HYPO71: A computer program for determining hypocenter, magnitude, and first motion pattern of local earthquakes: U.S. Geological Survey. Open-file Report, p. 100.

[9]. Lee, W. H. K., Bennett, R. E., and Meagher, K. L. (1972a), A method of estimating magnitude of local earthquakes from signal duration: U.S. Geological Survey. Open-file Report, pp 28.

[10]. Louie, J. (1996a), Seismic Deformation. URL: http://www.seismo.unr.edu/ftp/pub/louie/class/100/seismic-waves.html (Accessed 1 July 2009).

[11]. Louie, J. (1996b). What is Ritchet magnitude? Nevada Seismological Laboratory, URL: http://www.seismo.unr.edu/ftp/pub/louie/class/100/magnitude.html (Accessed 1 July 2009).

[12]. McEvilly, T.V. (1976), Seismological Instrumentation, in Seismic Risk and Engineering Decisions, E. Rosenblueth and C. Lomnitz (eds.), Elsevier, pp. 381-414.

[13]. McEvilly, T.V. (1982), Seismographic instrumentation, in Encyclopedia of Science and Technology, McGraw-Hill, New York, 5th ed.

[14]. Mcwuilin, R., Bacon, M., and Barclay, W., (1980), An Introduction to Seismic Interpretation, Graham & Trotman, London, UK.

[15]. Richter, C. F. (1958), Elementary Seismology, N H Freeman & Co, California, USA.

[16]. Savarensky, E. (1975), Seismic waves, Mir Publishers, Moscow.

[17]. Simon, R.B. (1981), Earthquake Interpretation: A manual for reading seismograms, Willian Kaufmann, Los Ator, California.

[18]. Trnkoczy, A. (1998), Understanding and setting STA/ LTA Trigger Algorithm parameters for the K2, Application Note 41 Rev A/97, Kinemetrics Inc USA, URL: http://www.kinemetrics.com/eng_ftp/AppNotes/appnote41.PDF.

[19]. IRIS (n.d), How are Earthquakes Located?, Education & Outreach Series No. 6, IRIS Consortium. URL: http://www.iris.edu/edu/onepagers/no6.pdf (Accessed 2 Dec 2006)

[20]. Klein, F. (n.d.), Finding an earthquake's location with modern seismic networks, Earthquake hazard Program-Northern California, US Geographical Survey USA. URL: http://quake.usgs.gov/info/eqlocation/index.html (Accessed 1 Sept 2005)

[21]. Attri, RK (2018/2001), A simplified Overview: How are the Earthquake Parameters Computed from the Recorded Seismic Signals?, R.Attri Instrumentation Design Series (Seismic), Paper No. 2, Seismic Instrumentation Design: Selected Research Papers on Basic Concepts, ISBN 978-981-11-9751-2, 2nd edn, pp. 21-34, Speed To Proficiency Research: S2Pro©, Singapore.

[22]. USGS (n.d.), How Do Seismologists Locate An Earthquake? - USGS, https://www.usgs.gov/faqs/how-do-seismologists-locate-earthquake (accessed December 26, 2018).

[23]. Aiken, C.L.V. (n.d.), Sound Waves And Seismic Waves [Lecture 6], University of Texas, Dallas, URL: http://www.utdallas.edu/~aiken/SHAKEBAKE/lecture6sp07.pdf (accessed December 27, 2018).

A SIMPLIFIED OVERVIEW: HOW ARE THE EARTHQUAKE PARAMETERS COMPUTED FROM THE RECORDED SEISMIC SIGNALS?

RAMAN K. ATTRI

Ex-Scientist (Geo-Scientific Instruments)
Central Scientific Instruments Organization, Chandigarh, India

Abstract: The earthquake phenomenon is one of the unpredictable and complex geological phenomenon happening under the earth surface. The recording and its interpretation of recorded earthquake signal have always been a prime concern of the seismologists. This paper discusses in a very simplified manner how these complex parameters are computed.

I.   INTRODUCTION TO SEISMOLOGICAL STUDIES

Due to sudden rupture within the earth, the accumulated strain energy is radiated in elastic or seismic waves. When these waves reach earth's surface, we feel them as `vibrations' and call the phenomenon - `earthquake'. Due to the Sudden Rupture, the Strain Energy accumulated in the earth is released abruptly in the form of elastic waves. These elastic waves are known as Seismic Waves. These are produced due to the fact that whenever there is a sudden Rupture, the strain energy accumulated in the earth is released abruptly in the form of elastic waves [1]. These seismic waves are generated naturally by an earthquake or artificially by an underground nuclear explosion or other means. These seismic waves originated from the source (focal point) travel inside the earth and finally reach the surface of the earth and are felt as vibrations. The total phenomenon is called 'Earthquake'. The Science dealing with such studies on strain wave propagation in the interior of the earth, interpretation of these seismic events and signals is called seismology. [2]

The seismological study is closely linked with implementing the right kind of the seismic instrumentation for recording these earthquakes, interpreting them, storing their history over the years. Software systems enhance the power of such instrumentation by performing the major job of signal analysis, complex computation of parameters, data interpretation, fetching inference, statistical trend analysis of seismic activity at a place of interest over the years. In association with instrumentation systems, the complex software system deployment sometime ensures accurate forecasting and generating warning systems for an earthquake. This makes the job of a seismologist more accurate, objective, automated and quick.

This paper presents the seismological interpretation process using software systems in a very simple manner. The approach has been described without going into the background of complex mathematically computations and complexity of software algorithms involved. Many equations and discussions unique to geology and earthquake sciences have been deliberately kept out of the scope of the paper to make it understandable to a wider segment of readers.

II. EARTHQUAKE MECHANISM: A BRIEF PREAMBLE

Much theory is available in the mechanism of the earthquake. The earth comprises several "shell" layers of which the outmost, hard layer - the lithosphere - consists of stiff plates. These plates are in constant motion, and as a result, stress builds up between them. This theory is known as plate tectonics [3, 4].

When an earthquake fault ruptures, it causes two types of deformation: static and dynamic. Static deformation is the permanent displacement of the ground due to the event. This type of deformation is explained with the help of the Elastic Rebound Theory [5]. The continuous motion of the earth plates causes stress to build up at the boundaries between the plates, where friction keeps the boundaries locked. Once the frictional stress (or the strength of the rocks) is exceeded, the two sides move (along a fault) causing an earthquake. This principle applies away from plate boundaries since the locked plate boundaries cause the stress to extend to the plate interior. Here, earthquakes occur where the local strength of the rock is less than the regional stress level. The key point of the Elastic Rebound Theory is that *the stress is continually building up* and that *earthquakes act to relieve that stress, as shown in figure 1.*

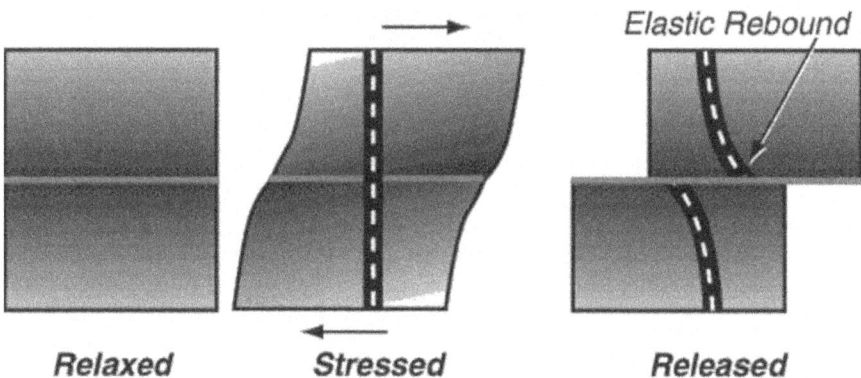

Fig. 1: Static Deformation due to Elastic Rebound [5]

Typically, someone will build a straight reference line such as a road, railroad, pole line, or fence line across the fault while it is in the pre-rupture stressed state. After the earthquake, the formerly straight line is distorted into a shape having increasing displacement near the fault, a process known as elastic rebound. For a constant rate of stress increase due to plate motion, the greater the time between earthquakes, the greater the stress release when the earthquake occurs (larger magnitude).

The friction causing the earthquake can be extensional, compressional or transformational depending upon the earth plate orientations [3, 4]. The typical scenarios are depicted in figure 2.

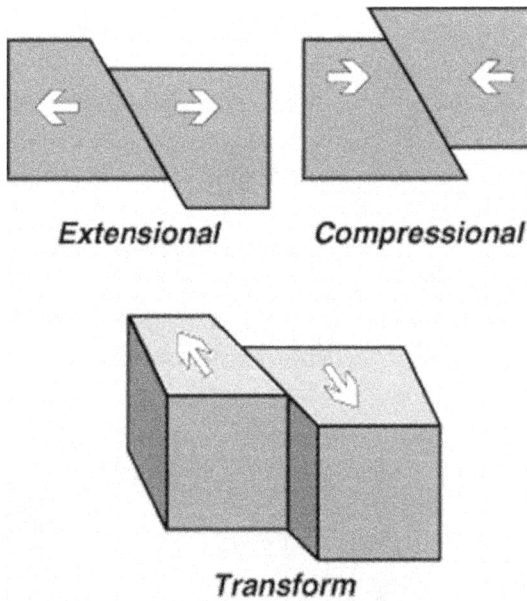

Fig. 2: *Extensional, Compressional or Transform frictional stress* [5]

Earthquakes generate seismic waves with a broad spectrum; that is, the waves can shake over a fairly broad range of frequencies. As these waves propagate away from the earthquake source, however, anelastic behavior of the rocks of the crust and uppermost mantle, cause the higher frequency components to be damped out. Thus, the farther a seismic wave propagates, the less high-frequency energy it will contain (this is called anelastic attenuation).

The second type of deformation, dynamic motions, is essentially sound waves radiated from the earthquake as it ruptures. While most of the plate-tectonic energy driving fault ruptures is taken up by static deformation, up to 10% may dissipate immediately in the form of seismic waves [3, 4].

Seismic waves are used in locating and modeling earthquakes and underground nuclear explosions, and for imaging the interior structure of the earth. The seismic waves have been designated as Body waves and Surface waves [2].

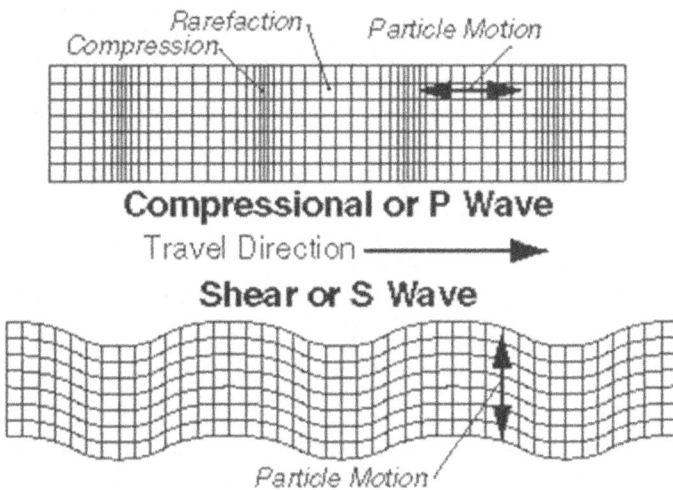

Fig. 3: Direction of particle motion in P- and S-wave [5]

Body waves are further classified as 'P' waves (Longitudinal wave) and 'S' waves (Transverse waves) [5]. The P (primary wave) is also called the compression wave traveling away from a seismic event through the earth's crust, and consisting of a train of compressions and dilatations of the material (push and pull). The S (secondary) wave is also called a shear wave, produced essentially by the shearing or tearing motions of earthquakes at right angles to the direction of wave propagation. The mechanical properties of the rocks that seismic waves travel through quickly organize the waves into these two types. P-waves travel fastest, at speeds between 1.5 and 8 kilometers per second in the earth's crust. S-waves, travel more slowly, usually at 60% to 70% of the speed of P-waves. P-waves shake the ground in the direction they are propagating, while S-waves shake perpendicularly or transverse to the direction of propagation, as shown in figure 3.

Other types of waves are Surface waves, which follow the earth's surface only, with speed less than that of S-waves. Surface waves are categorized as Rayleigh waves and Love waves. Love wave produces a sideways motion, whereas Reyleigh waves are forward and elliptical vertical seismic surface waves [2, 6].

The earthquake and its mechanism are understood to a greater extent by studying the properties of these waves using Instruments. The Instrumental Measurement and Analysis of seismic waves is called Seismometry. The instruments generally record the P-wave and S-wave since these waves are travel faster through the earth crust and can be captured at various remote data collection points.

As for earthquake effects, they may vary depending upon the distance of the earthquake occurrence and the structure under question. For example, the response of a building to shaking at its base due to seismic waves depends on several factors related to its design and construction [7]. However, one of the most important factors is simply the height of the building, because this determines the frequency of resonance of the building. Short buildings have a high resonant frequency (short wavelength), while tall buildings have a low resonant frequency (long wavelength). In terms of seismic hazard, therefore, short buildings are susceptible to damage from high-frequency seismic waves from relatively near earthquakes. But tall buildings are at risk due to low-frequency seismic waves, which may have originated at a much greater distance.

III. INSTRUMENTATION FOR RECORDING OF SEISMIC WAVES

Unfortunately, the earth is not transparent, and we cannot just see or photograph the earthquake disturbance like meteorologists can photograph clouds [8]. When an earthquake occurs, it generates an expanding wavefront from the earthquake hypocenter at a speed of several kilometers per second [23]. Generally, it takes few second for theses waves front to travel thousands of kilometers [8]. A wavefront expansion is shown in figure 4.

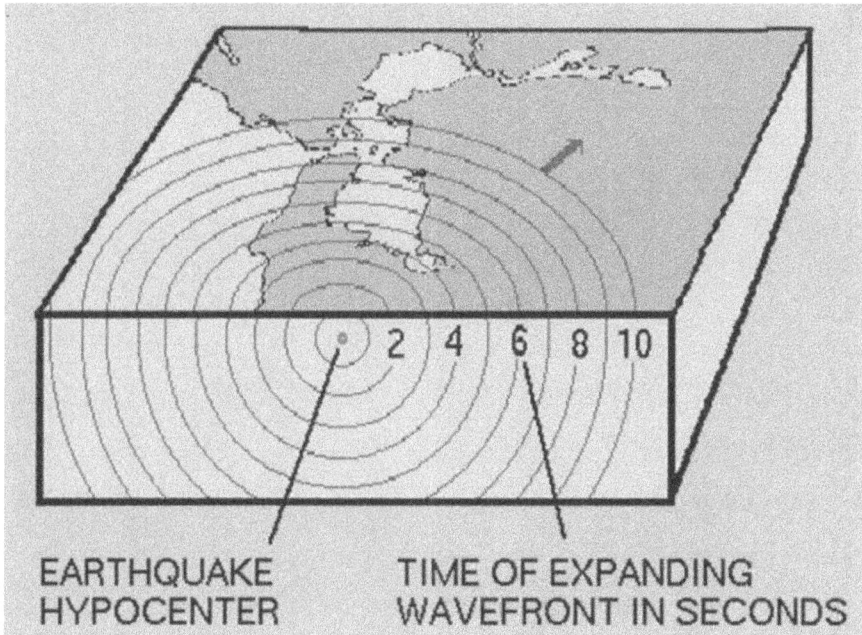

Fig. 4: Wavefront expansion from hypocenter [8]

This wavefront consists of two unique waves: One P-wave which comes earlier due to its faster speed and another S-wave front which comes later due to the little slower speed of travel. As shown in figure 5, the P-wave front is released earth by the earthquake and thus reached the seismic station (shown as 'A') and S-wave front soon follow the P-wave front. Thus, we get two unique wave peaks on the recording instruments, each corresponding to it wavefront.

We observe earthquakes with a network of seismometers on the earth's surface. The ground motion at each seismometer is amplified and recorded electronically at a central recording site. As the wavefront expands from the earthquake, it reaches more distant seismic stations [24]. More than one sensor or seismic stations, minimum three stations are required to pinpoint the location of the earthquake rupture.

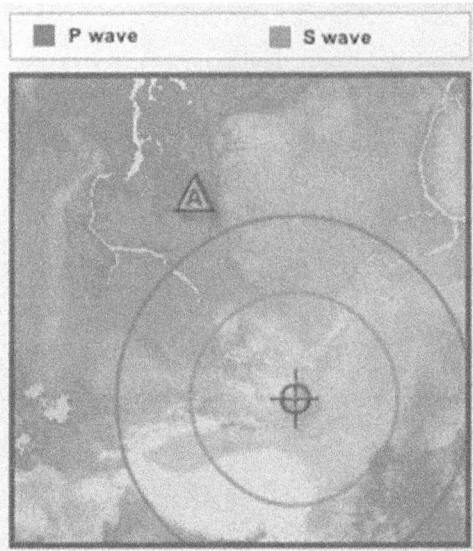

Fig. 5: S-wave front following the P-wave front in an actual earthquake

A seismograph is the main instrument to detect measure & analyze the seismic signal emanated either from a natural source (earthquake) or due to an artificial cause (underground nuclear explosion/ blast). With developing advanced measurement techniques and new technologies, seismographs have also progressed into several phases of technology up gradation starting from simple analog recorder type to one based on a microprocessor or high-performance built-in PC. Modern seismic stations have very sensitive and sophisticated sensors and data acquisition systems. [9]

The development in seismological instruments has closely followed the ongoing advancement in technology and the progress made on understanding the Physics of Earthquake. The first device for measuring earthquake was seismoscope. It was made by Dr AD Cheng of China in 132 AD. The quantitative measurement of an earthquake by using instrument came in 1880 in Japan whereas Teleseismic Recording of an earthquake was taken in 1890. The Wood-Anderson Torsion Seismograph was introduced in 1922 and is still in use. In the latter part of this century, appeared the Electromagnetic Seismograph which provided much higher magnification and provided the basis of modern seismographic instrumentation.

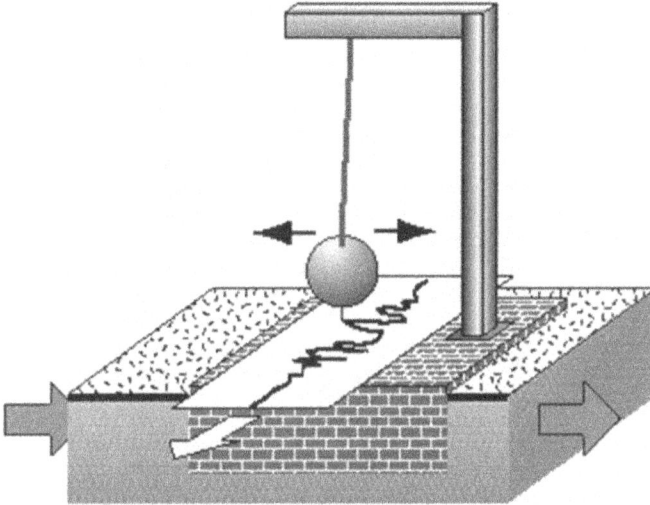

Fig. 6: Basic suspended pendulum Seismograph [5]

The principle behind earthquake recording is very simple. In the basic kind, it could be a suspended pendulum tied with a marker, as shown in figure 6. This was the model which used to be used in earlier years. However, this model clarifies it the concept behind the seismic instrumentation [5, 14].

In the modern instruments, the earthquake vibrations are generally sensed by an electromagnetic sensor. This sensor is a coil suspended in a magnetic field. The vibration induces EMF in the coil proportional to the amplitude of the vibration, and the characteristics of this electrical signal contain the characteristics of the earthquake like frequency, amplitude, direction, p and s-waves. This electrical signal is stored in electronic form. The storage could be on an analog smoke paper, the thermal plotter, digital cassette or solid-state memory or PC hard-disk. [10, 14] The underlying idea is to record the vibration impressions on paper in the form of electronics or no-electronics signal waveform, typically as shown in figure 7 which is the waveform recorded on an Analog Seismograph on a thermal paper [11, 14].

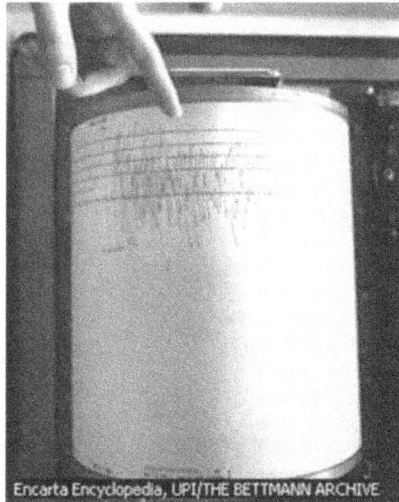

Fig. 7: Analog Seismogram taken on a thermal paper [11]

The seismic signals recorded by these instruments may be in the form of a waveform on a paper, or it could be an electronic data. A typical waveform recorded on a true earthquake with the help of seismic sensor is shown in figure 8 which shows two peaks corresponding to P-wave and S-wave respectively.

Fig. 8: A typical seismic signal during an earthquake [12]

It can be seen that the signal contains a lot of residual background noise and high peaks of the earthquake event. Due to such a scenario, there are two requirements on the sensor. One, that sensor must be very sensitive. The reproduced electrical signals exhibit an extremely low amplitude of (nv to mv). Two, the sensor must exhibit a large dynamic range of the sensor since the dynamic range of the earthquake can be very large. Sometimes as dynamic range as wide as 200 dB (ten order of magnitude) is resulted due to Ground Displacement as small as few nanometers to as large as several Meters at the epicenter of a major earthquake. Normally an amplifier with a wide dynamic gain/range (> 120 dB) is required to amplify the signal from the sensor faithfully. Further, these signals are highly noise corrupted signals. To handle such a complex signal, one needs extraordinarily precise Instruments which can distinguish between the extremely feeble signal and dominant background noise. Signal Bandwidth covers a range from nearly DC to over 100 Hz for earthquakes, mostly in the band (0.001 Hz - 100 Hz). Since no single instrument can operate over such a wide bandwidth and dynamic range, therefore, a set of instruments for different bandwidth have come up [9].

Seismic data obtained from must be correlated. Seismic Networks Arrays hooked up around this instrument must be capable of detecting true seismic events, producing good quality digital data in enormous quantity from the Multiple Sensors in parallel, Processing and Transmitting the data further analysis and interpretation.

Specific kind of instruments and systems are required to the feeble signals the dominant background noise. These instruments/ systems must generate electrical signals corresponding to the change in variables, produce digital data in copious quantity from multiple sensors in parallel and finally Process, compress and transmit it for further analysis and interpretation to evaluate earthquake parameters such as—

- Time of occurrence of an event
- It's epicenter (point exactly above the focal point)
- The focal point (depth of source)
- Its magnitude and polarity (P and S-wave)
- Its coda length (total-event duration)
- It's nature (source - discrimination)

IV. PARAMETERS OF SEISMIC VIBRATIONS

An earthquake is a unique form of vibration taking place inside the earth surface. Few characteristics are of the interest to a seismologist [13]. These include:

a) Where did the earthquake generate and what was its direction of propagation? Location & Direction
b) When did the earthquake occur? Time
c) What was the ground motion? Acceleration
d) What was the magnitude of the earthquake? Magnitude

Other important question could be: What is the general seismic activity of that area?

A. *Location Parameters*

The epicenter is an important term in seismology; this is the point exactly above the earthquake on the earth surface.

The point where the earthquake generates is called hypocenter. Hypocenter is also called the focal point sometimes. Regarding epicenter and hypocenter, these types of distances are computed, which provide a good amount of information to the seismologist. These distances are graphically shown in figure 9.

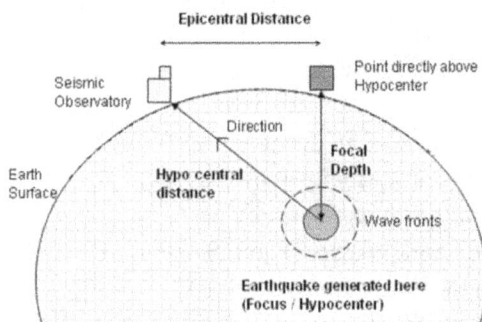

Fig. 9: Graphical view of Earthquake Locational Parameters

- o **Focal Depth:** The vertical distance from the focal point to the epicenter (i.e. point exactly above it vertically on the earth surface) is called Focal Depth. This gives the depth of the focus or hypocenter beneath the earth's surface. It commonly classes earthquakes like shallow (0-70 kilometers), intermediate (70-300 kilometers), and deep (300-700 kilometers). [14]

- o **Hypocenter Distance:** The earthquake creates its effects in the areas not just above its location; instead it also affects the areas falling on its vector direction. A distance "Hypocenter Distance" is defined to indicate the distance between focus or hypocenter and the point of observation or the point where an earthquake has produced its effect. This is a straight-line distance. [14]

- o **Epi-central Distance:** This is the straight-line distance between the point of observation of an earthquake and the point exactly above the focal point on the earth surface. This is measured in degrees and can be converted in kilometers depending upon the parameter constants of the location where the earthquake is being measured. [14]

B. Timing Parameters

The timing parameters in seismology are taken regarding the arrival of P- and S-wave. Generally, there are two types of timing parameters namely: S-P time interval and Coda Length of prime importance to seismologists. These are depicted in figure 10.

The S-P time interval: This the difference between the arrival time of P and arrival time of S. Most of the earthquakes are recorded with timing information so it is quite easy to record this information. This time gives the estimate about the seismic activity and profile of the site and nature of an earthquake. How closely S and P-waves follow each other provides important inferences regarding the creation of some early warning systems. The seismologists analyze the S-P time difference data over the years for a particular location and can come up with minimum time difference available between reception of P-wave and actual heavy disastrous S-wave. This information may be used for generating quick earthquake warning alarms in the locations of strategic importance.

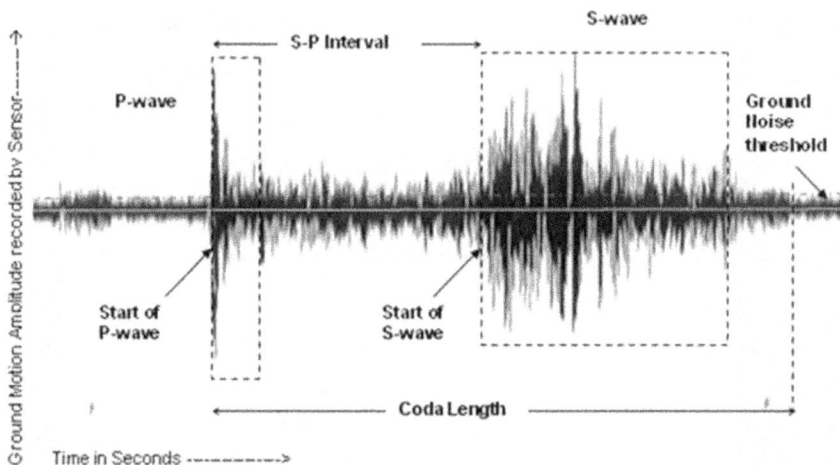

Fig. 10: Time Parameters of a seismic signal

Since P-waves travel faster than S-waves, the time difference between the arrival of the P-wave and the arrival of the S-wave depends on the distance the waves traveled from the source (earthquake) to the station (seismograph). Over time, many such measurements have been made, and travel-time curves (time vs. distance plots for P, S and other more complicated waves) have been developed for the earth. If an earthquake occurred a given distance from a station, it could have been anywhere on a circle whose radius is that distance, centered on the station. If distances from two stations are known, two locations are possible: the two intersection points of two circles. If distances from three stations are known, the earthquake can be unambiguously located. This is the principle of triangulation.

So, normally we know the velocity of P & S-waves. We can calculate the Epicentral distance Δ in kms from the S-P interval as under:

$$Epicentral\ Distance\ in\ Kms,\ \Delta = V_p / (V_p/V_s\text{-}1)\ (T_s\text{-}\ T_p)\quad \text{-eq (1)}$$

Where V_p and V_s are the velocities of P-wave and S-wave respectively and T_p and T_s are their respective arrival times.

For crustal rocks, typical V_p is 6 km/s and V_p/V_s is approximately 1.8

km. So Epicentral distance in kilometers is about 7.5 times S-P interval in seconds. If three or more Epicentral distances measure by various seismic stations are available, it is possible to use a map to plot these data. The epicenter may be placed at the intersection of circles with stations as a center and appropriate Δ as radii. This point of intersection may be an aerial distance giving the uncertainty in locating exact epicenter. This will be clearer in the following section addressing "locating earthquake using software approach."

Coda Length: This is basically the time in Seconds starting from P-wave time to the end of a wave up to noise level. This time indicates how much time seismic shock prevailed and how much time was taken by the earth to calm down after the shock. This time is very important is relating the amount of destruction with the time interval of the earthquake. This time measurement starts from right from the occurrence of the P-wave. Roughly it is the duration of a true seismic event, so it may also be referred as total signal duration. For a given earthquake, coda length is measured by different stations and an average value of these durations is taken [14].

C. *Ground Motion*

To characterize or measure the effect of an earthquake on the ground (a.k.a. ground motion), the ground motion parameters are sometimes used. The most important is the acceleration of the ground motion. *Acceleration* is the rate of change of speed, measured in "g"s at 980 cm/sec^2 or 1.00 g. For example, 0.001g or 1 cm/sec^2 is perceptible by people, 0.02 g or 20 cm/sec^2 causes people to lose their balance, 0.50g is very high but buildings can survive it if the duration is short and if the mass and configuration has enough damping [15].

Another parameter is *Velocity* (or speed) which is the rate of change of position, measured in centimeters and *Displacement* is the distance from the point of rest, measured in centimeters [15].

D. *Magnitude Parameters*

Each earthquake has a unique amount of energy, but magnitude values given by different seismological observatories for an event may vary.

Depending on the size, nature, and location of an earthquake, seismologists use several methods to estimate magnitude [5]. The seismologists often revise magnitude estimates as they obtain and analyze additional data. Although each earthquake has a unique Magnitude, its effects will vary greatly according to distance, ground conditions, construction standards, and other factors [5]. For this purpose, seismologists use a different Mercalli Intensity Scale to express the variable effects of an earthquake. We can describe this in two ways:

- The magnitude on the Richter scale; this measure relates to the energy released by the earthquake, and it is independent of possible damages on the surface of the earth.

- The intensity, which is a measure of the shaking at the earth's surface. Here the Modified Mercalli scale is used.

A relatively small magnitude earthquake that happens near the surface can cause shaking of great intensity, whereas a large magnitude earthquake that happens in a depth of several hundred kilometers will not necessarily produce intense shaking at the surface.

Here we will concentrate only on traditional computation-based scales to find the magnitude of the earthquake. It is popularly measured with the help of three scales.

Ritchet Scale Magnitude: This a universal scale specified by the Ritchet to measure the intensity of the earthquake. Size of the earthquake in Richter scale is given on a 10-point scale which ranges from 1-10[16]. All the entire earthquakes across the world are measured on this scale to keep the uniformity in the comparisons.

Ritchet scale is normally determined based on maximum amplitude shown on Wood-Anderson seismograph (which have nearly constant displacement amplification over the frequency range of local earthquake). Ritchet scale defined the Local Earthquake Magnitude, observed at the place of observation, denoted by M_L irrespective of the direction and its origin and empirically given as under:

$$M_L = Log\ A - Log\ A_0\ (\Delta) \hspace{4cm} \text{-eq (2)}$$

Where A= Maximum amplified in mm on Wood-Anderson seismographs, Δ is the Epicentral distance in kilometers and A_O (Δ) is maximum amplitude at Δ kms for a standard earthquake. Local Magnitude is thus a number characteristic of earthquake and independent of the location of the recording seismographs. The second factor was scaled by certain assumptions by Ritchet. To solve the above equation (2), a table of $-\log A_O$ as a function of Epicentral distance in kilometers is needed. Ritchet arbitrarily chose $-\log A_O = 3$ at $\Delta = 100$ kms and other entries in the table were constructed from observed amplitudes of a series of well-located earthquakes. [19]

Beno Getenberg & Chrles F. Ritchet (1940), extended local scale magnitude of distant earthquakes. It was called Surface Wave Magnitude, denoted by M_S and empirically given as under:

$$M_S = Log\ A - log\ A_O\ (\Delta_O) \qquad\qquad \text{-eq (3)}$$

Where Δ_O is epicentral distance taken in degrees to consider and origin of the origin of the earthquake, A is maximum combined horizontal ground amplitude in micrometers for surface waves with a period of 20 sec and $-\log A_O$ is tabulated as a function of epicentral distance Δ in degrees, similar to that for local magnitude.

However, this applied to shallow earthquakes generating observable surface waves. However, for distant earthquakes at various depths inside the earth, only body waves are observable as P and S-waves. So the magnitude had to be defined based on observed P-wave and S-wave amplitudes. A new relationship for body waves was defined as Body Wave Magnitude, denoted as M_B and empirically given as under:

$$M_B = log\ (A/T) - f\ (\Delta.\ h) \qquad\qquad \text{-eq (4)}$$

Here (A / T) is the maximum amplitude to period ratio in micrometer per second, f $(\Delta,\ h)$ is a calibrated function of Δ and focal depth h. This function has been characterized for most major locations across the globe and empirically it can be written as:

$$M_B = log\ (A/T) + \Delta + C \qquad\qquad \text{-eq (5)}$$

Where A is the amplitude of the highest S-wave Peak, T is the time period of the same peak, as shown in figure 11.

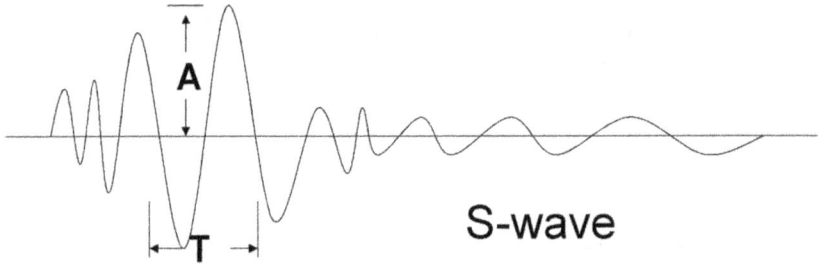

Fig. 11: Parameters required in the seismic wave for Ritchet scale calculation

C is place specific constant. This *distance factor* comes from a table that can be found in Richter's (1958) book *Elementary Seismology* [17]. Theoretically, it is the logarithmic value of the amplitude of the highest peak with the time period of that peak. Higher the amplitude, higher the strength of the earthquake and higher the time period lower will be the strength. This is interpreted like this that high peaks which are sharpest most indicate the highest strength of the earthquake occurrence.

Seismologists use a Ritchet Magnitude scale to express the seismic energy released by each earthquake. Table [1] lists the typical effects of earthquakes in various magnitude ranges.

Table [1]: Relationship of Ritchet Magnitude with effects [5]

Richter Magnitudes	Earthquake Effects
Less than 3.5	Generally, not felt, but recorded.
3.5-5.4	Often felt, but rarely causes damage.
Under 6.0	At most slight damage to well-designed buildings over small regions.
6.1-6.9	Can be destructive across where people live in areas up to about 100 kms.

| 7.0-7.9 | Major earthquake. Can cause serious damage over larger areas. |
| 8 or greater | Great Earthquake. Can cause serious damage in areas several hundred kilometers across. |

Coda-length Magnitude: It is an estimate of local magnitude M_L calculated using the coda-length/magnitude relationship [14]. Coda Magnitude is directly related to the coda length and epicentral distance. This approach is based on using the signal duration rather than the maximum amplitude to estimate the earthquake magnitude because it will make it independent of necessarily using Wood-Anderson Seismograph. E. Bisztricsany (1958), SL. Solvov (1965), W.H.K. Lee (1972) correlated Ritchet magnitude with a signal duration of the local earthquake through an empirical formula as under:

$$M_D = -0.87 + 2 \log D + 0.0035 \, \Delta \qquad \text{-eq (6)}$$

Where D   =   Coda length
Δ   =   Epicentral Distance

Similar empirical relationships in many forms integrating many dependent factors have been suggested by many researchers, the more general form of the same is as under:

$$M_D = a_1 + a_2 \log D + a_3 \, \Delta + a_4 \, h \qquad \text{-eq (7)}$$

Where h is the focal depth and a_1, a_2, a_3 are empirical constants. The above equation can be empirically established for a location of interest depending upon certain empirical relationship. For example, the generalized empirical relationship reduces to following for a city like Chandigarh as under:

$$M_D = -0.57 + 1.38 \log D + 0.29 \, \Delta \qquad \text{-eq (8)}$$

The underlying concept is that coda magnitude is a measure of the impact of an earthquake which occurs at a specified epicenter in a particular direction, specified epicentral distance and sustains for a specified duration in seconds.

Seismic Energy: Both the magnitude and the seismic moment are related to the amount of energy that is radiated by an earthquake. Richter, working with Dr. Beno Gutenberg, early on developed a relationship between magnitude and energy. Their relationship is [5]:

$$Log\ E_S = 11.8 + 1.5 M_S \qquad\qquad -eq\ (9)$$

giving the energy E_S in ergs from the surface wave magnitude M_s. Note that E_S is not the total "intrinsic" energy of the earthquake, transferred from sources such as gravitational energy or to sinks such as heat energy. It is only the amount radiated from the earthquake as seismic waves, which ought to be a small fraction of the total energy transferred during the earthquake process [18]. The Richter scale is based on the maximum amplitude of certain seismic waves, and seismologists estimate that each unit of the Richter scale is 31 times increase of energy [15].

V. LOCATING EARTHQUAKE USING SOFTWARE APPROACH

The earthquake is measured by a network of seismic stations located at different locations. When an earthquake occurs, we observe the times at which the wavefront passes each station. We find the unknown earthquake source by knowing these wave arrival times at different seismic stations. [5]

We can locate earthquakes using a simple fact: an earthquake creates different seismic waves (P-waves, S-waves, etc.) The different waves each travel at different speeds and therefore arrive at a seismic station at different times. P-waves travel the fastest, so they arrive first. S-waves, which travel at about half the speed of P-waves, arrive later. A seismic station which is close to the earthquake records P-waves and S-waves in quick succession. With increasing distance from the earthquake, the time difference between the arrival of the P- waves and the arrival of the S-waves increases. The time interval can find the location of the earthquake.

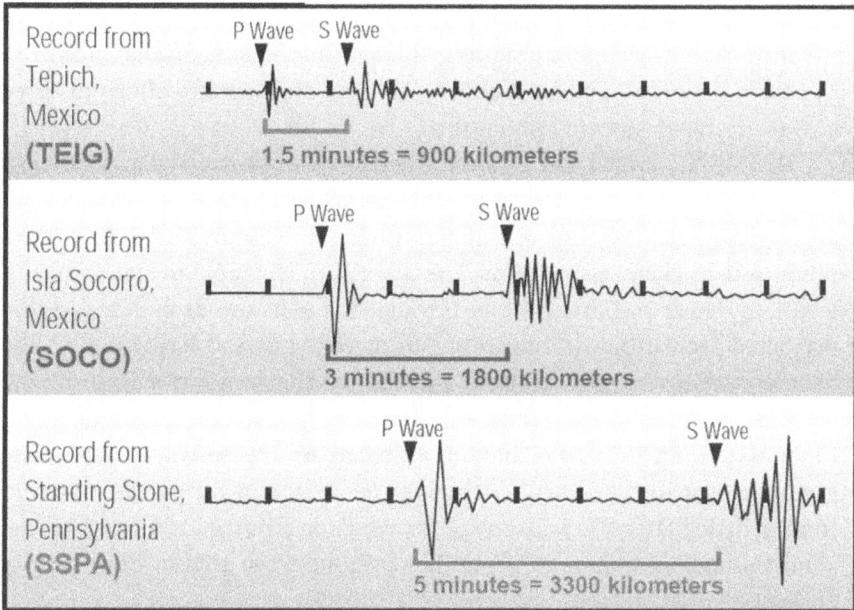

Fig. 13: The seismic signal recorded at three stations from the same earthquake (Courtesy: IRIS Consortium & US Geological Survey) [18]

As a simple illustration, we present here a real data of earthquake occurred in Mexico (Courtesy: *IRIS Consortium - University of Arizona, University of California, Berkeley, University of California, San Diego, Purdue University, and the US Geological Survey)* [18]. The seismic signal at three locations was recorded as shown in figure 13. The arrival time of P-wave and S-wave and S-P interval is different at three locations depending upon the distance from the earthquake.

The time taken by P and S-wave is different at each station and further, the interval depends also upon the distance of the station from the location of the earthquake. Based on S-P interval, Epicentral distance is calculated.

From observing and analyzing many earthquakes, we know the relationship between the S-P time and the distance between the station and the earthquake. We can, therefore, convert each measured S-P time to distance. A time interval of 1.5 minutes corresponds to a distance of 900 kilometers, 3 minutes to 1800 kilometers, and 5 minutes to 3300 kilometers. Once the distance to the earthquake for three stations is known, the location of the earthquake can be determined. For each station, we draw a circle around the station with a radius equal to its distance from the earthquake [18]. The earthquake occurred at the point where all three circles intersect, as found in figure 14.

A straightforward and general procedure can be evolved and can also be programmed in a computer with which we can find the location, depth and origin time of an earthquake whose waves arrive at the times measured on each seismogram. The algorithm is simple to state: guess a location, depth and origin time; through the software itself compare the predicted arrival times of the wave from your guessed location with the observed times at each station; then move the location a little in the direction that reduces the difference between the observed and calculated times [24]. Then this procedure is repeated by the software, each time getting closer to the actual earthquake location and fitting the observed times a little better [24]. Software stops when its adjustments have become small enough and when the fit to the observed wave arrival times is close enough [8].

Mathematically, the problem is solved by setting up a system of linear equations, one for each station. The equations express the difference between the observed arrival times, and those calculated from the previous (or initial) hypocenter, in terms of small steps in the 3 hypocentral coordinates and the origin time. We must also have a mathematical model of the crustal velocities (in kilometers per second) under the seismic network to calculate the travel times of waves from an earthquake at a given depth to a station at a given distance. The system of linear equations is solved by the method of least squares which minimizes the sum of the squares of the differences between the observed and calculated arrival times [24]. The process begins with an initially guessed hypocenter, performs several hypocentral adjustments each found by a least squares solution to the equations, and iterates to a hypocenter that best fits the observed set of wave arrival times at the stations of the seismic network [8].

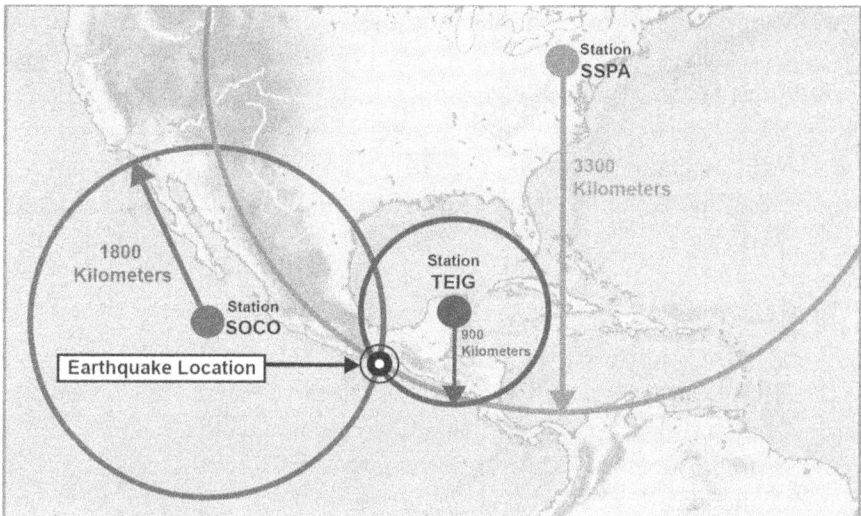

Fig. 14: Method of Circle intersection to find earthquake location (Courtesy: IRIS Consortium - the University of Arizona, University of California, Berkeley, University of California, San Diego, Purdue University, and the US Geological Survey) [18]

VI. CONCLUSION

Now interdisciplinary approaches are being taken to use the seismology to understand seismic profile of the site, general noise level and earth activity at the desired site, trends of events occurring there, finding a clue to successful prediction of earthquake occurrence, generating suitable warning signal in time, protecting man-kind and property by taking timely actions [20, 21]. The powerful software tools exist which analyzes the site and provide a 3-D profile of the impact of the earthquake. The software for earthquake analysis has reached it a state-of-the-art position where very fast and accurate earthquake analysis archived over the years is possible, and the results are being used for creating an effective earthquake forecasting mitigation and warning system [22]. However, the prediction and forecasting of an earthquake are still is a science of great interest and significant research, modeling and techniques are being evolved continuously. The prediction process integrates the sophistication of seismic instrumentation, networking of seismic stations on a global scale, an exhaustive integrated database of a range of earthquakes across the globe, powerful seismic signal processing capabilities, state-of-the-art signal analysis software, virtual modeling, statistical data analysis and telecommunication coupled warning systems. The approaches are underway and hopefully will get formalized in two years to save humanity on a mass level.

ACKNOWLEDGMENTS

1) Geo-Scientific Instruments Division, Central Scientific Instruments Organization, Chandigarh, INDIA, www.csio.nic.in
2) Naresh Kumar, M.Sc (Seismology), Ex-Seismologist, CSIO, Chandigarh
3) Earthquake Hazard Program, US Geographical Survey, USA, http://wwwneic.cr.usgs.gov/
4) IRIS, www.iris.org
5) Seismolink, http://www.seismolinks.com/
6) Encyclopedia of Physical Science & Technology, Vol 12, Academic Press, Inc, 1987.

VII. REFERENCES

[1]. Macelwane, J.B, & Sshon, F.W. (1932), Introduction to Theoretical Seismology Part-I, Geodynimics, St Louis University, Saint Louis, Missouri, USA.

[2]. Madariaga, R. (1987), Theoretical Seismology, Encyclopedia of Physical Science & Technology, Vol 4.

[3]. Louie, J. (2001), Plate Tectonics, the Cause of Earthquakes, URL: http://www.seismo.unr.edu/ftp/pub/louie/class/100/plate-tectonics.html

[4]. Kasahara, K. (1981), Earthquake Mechanics, Cambridge University Press, London.

[5]. Louie, J. (1996) Seismic Deformation, URL: http://www.seismo.unr.edu/ftp/pub/louie/class/100/seismic-waves.html.

[6]. Savarensky, E. (1975), Seismic waves, Mir Publishers, Moscow.

[7]. US Army Corps of Engineers (1998), Seismic Design For Buildings, , Engineering Division, Directorate of Military Programs, Washington, URL: http://www.ccb.org/docs/COETI/ti809_04.pdf.

[8]. Klein, F. (n.d.), Finding an earthquake's location with modern seismic networks, Earthquake hazard Program-Northern California, US Geographical Survey, USA, URL: http://quake.usgs.gov/info/eqlocation/index.html.

[9]. Thomas, V. (n.d.), Seismological Instrumentation, Dept of Geology and Geophysics, University of California, California, USA.

[10]. Ambuder, B.P. and Soloman, S.C. (1974), An event recording system for monitoring small earthquakes, Bulletin of Seismological Society of America, Vol 64, pp 1181-1188.

[11]. Mirosoft (2000), Earthquake, Microsoft Encyclopedia Encarta-Standard Edition, Microsoft Corp.

[12]. Kumar, S, Attri, R.K., Sharma, B.K., Shamshi, M.A. (2000), Software Tools for Seismic Signal Analysis, IETE Journal of Education, Vol 41, No. 1& 2, pp 23-30, URL: https://doi.org/10.1080/09747338.2000.11415718.

[13]. Mcwuilin, R. et. al (1980), An Introduction to Seismic Interpretation, Grahans & Trotman Ltd, UK.

[14]. Simon, R.B. (1981), Earthquake Interpretation: A manual for reading seismograms, Willian Kaufmann, Los Ator, California.

[15]. Lorant, G. (2005), Seismic Design Principles, Whole Building Design Guide, Lorant Group, Inc. & Gabor Lorant Architects, Inc., URL: http://www.wbdg.org/design/seismic_design.php.

[16]. Louie, J. (2001), What is Ritchet magnitude, Nevada Seismological Laboratory, URL: http://www.seismo.unr.edu/ftp/pub/louie/class/100/magnitude.html.

[17]. Richter, C.F. (1958), Elementary Seismology, W H Freeman & Co (Sd), California, USA.

[18]. IRIS (n.d.), How are Earthquakes Located? IRIS Consortium, URL: http://www.iris.edu/edu/onepagers/no6.pdf.

[19]. Harold, J. & Bullen, K.E. (1948), Seismological Tables, Office of the British Association, W.I. London, UK.

[20]. Rikitake, T. (1976), Earthquake Prediction, Else-vier, Amsterdam.

[21]. Rikitake, T (1982), Earthquake Prediction, Encyclopedia of Physical Science & Technology, Vol 4.

[22]. UNESCO (1984), Earthquake Prediction, Proceeding of International Symposium on Earthquake Prediction, Paris.

[23]. Attri, RK (2018/2005), Backend Framework and Software Approach to Compute Earthquake Parameters from Signals Recorded by Seismic Instrumentation System, R. Attri Instrumentation Design Series (Seismic), Paper No. 3, Seismic Instrumentation Design: Selected Research Papers on Basic Concepts, ISBN 978-981-11-9751-2, 2nd edn, pp. 1-20, Speed To Proficiency Research: S2Pro©, Singapore.

[24]. USGS (n.d.), How Do Seismologists Locate An Earthquake?, USGS, https://www.usgs.gov/faqs/how-do-seismologists-locate-earthquake.

FURTHER READINGS

- Jeffrey S. Barker, Demonstrations of Geophysical Principles Applicable to the Properties and Processes of the Earth's Interior, <u>NE Section GSA Meeting</u>, Binghamton, NY, March 28-30, 1994.

- Teacher's Web resources on Web, URL: http://www.ees.nmt.edu/Geop/NMPEPP/scream/htmlli1.html

- Redwood City Public Seismic Network, Redwood City, California USA, http://psn.quake.net/

- Nevada Seismological Observatory, Earthquake Information Centre, URL: http://www.seismo.unr.edu/htdocs/abouteq.html

- Glossary of Seismic Terminology, <u>Whole Building Design Guide,</u> URL: http://www.wbdg.org/pdfs/seismic_glossary.pdf

- Seismogram Analysis, <u>Training Outline</u>, geotechnical Corporation, Texas, USA, April 1960, http://psn.quake.net/info/analysis.pdf

- Observational Seismology/ Theoretical Seismology, Earthquake Prediction, <u>Ency of Phys Sciences & Tech</u>, Vol 4 & 12, Academic Press, Inc, USA, 1987

GSIS: A CONCEPTUAL MODEL FOR WEB-BASED INTEGRATION OF INFORMATION TECHNOLOGY WITH GEOSCIENTIFIC INSTRUMENTATION

RAMAN K. ATTRI

Ex-Scientist (Geo-Scientific Instruments)
Central Scientific Instruments Organization, Chandigarh, India

Abstract: Since a long time, leading organizations have been advocating the need for implementation of IT in a centralized manner on the distributed geological data monitoring centers. In the era of IT, it is highly important to access, share and analyze related data on a common platform for fast retrieval of information in centralized and in distributed mode. In this article, a concept to integrate Geo-Scientific Instrumentation systems to information technology has been highlighted. The vision to create a Geo-Scientific Information System (GSIS) would be an approach where a centralized web-enabled database of the data acquired & transmitted by geographically distributed but networked observatories will be maintained and accessible over the WAN. GSIS couples together the virtually independent domains of instrumentation, web-based data acquisition, database, IT, networking, telecommunication & internet together to enable users to access and retrieve information from sophisticated over the internet.

I. INTRODUCTION

The field of instrumentation mainly deals with the measurement of physical parameters in terms of some variables, which are finally converted into data and information is extracted from this data. Instrumentation Technology has reached to a level where great reliability in measurements of highly complex geological parameters can be expected. Geological & geotechnical instrumentation is a key area concerned with mankind safety. It ranges from earthquake prediction to glacier landslides forecasting. Geological observatories equipped with the latest state-of-the-art instrumentation systems are often the only way to obtain the needed information to understand individual geological manifestations like an earthquake, volcano, snow hydrological and oceanographic changes. These geographically distributed observatories measure the raw parameters and the after processing; final data gives the related geophysical variables, which helps in the study, forecast, and assessment of possible effects caused by this natural phenomenon.

Generally, each observatory transmits raw/processed data to a central station, and a database is maintained there. Real-Time monitoring systems and centralized data collection platform integrating distributed observatories measuring few parameters do exist in some countries, which provide online access to measured data only to restricted user segments only. This diversely distributed, but highly important data about resource and humanity safety must be provided to the users in another part of the world for online sharing and instant analysis via a universally accepted mode of information presentation such as the internet. A most important aspect of data (recorded by these geoscientific instrumentation systems) is the extraction of useful information and providing it to users all over the world in under some protocol of understanding.

In this view, a system concept has been proposed here to implement information technology revolution into the field of geo-scientific measurements resulting in a new area of Geo-Scientific Information System (GSIS). The concept is to establish a network of distributed geo-instrumentation platforms, which produces the necessary observational data. Then this data is logged onto the Main Server through local, regional and national telemetry stations. Main server centrally integrates, share and analyze the distributed data, and present the processed information for online access via the internet to users & experts. The system aims to assist scientists and users with both data and information solutions.

GSIS can be applied to different geological and geo-scientific studies such as earthquake studies, oceanographic studies, climatic studies, volcano eruption assessment, and snow hydrological studies. The individual phenomenon will have it own dedicated GSIS system serving altogether different user community. Here we shall take an example of seismic distributed instrumentation, one of the important sciences meant for human safety. It calls upon heavily the need to integrate sophisticated information, computer and communication technologies together.

In this paper, the design concept, vision & methodology to integrate existing Seismic Instrumentation systems to the information technology is discussed. Further, the dependencies of various sub-systems have been highlighted.

II. EXISTING SEISMIC MONITORING SYSTEM

Existing distributed seismic monitoring system consists of "nodes" and one or all of the "local seismic station," "regional seismic station", "national or central seismic station." The configuration depends upon the area being covered and the scope of the instrumentation. For example, for dam safety instrumentation, only one "local seismic station" is sufficed whereas earthquake mitigation plan may require "national seismic station" along with many local and regional stations. To accommodate everything, we will make our study in a later scenario.

Processor controlled configurable data acquisition systems in conjunction with the seismic sensors have been used in the seismic observatories all over the world for monitoring of seismic activities. Most of these instruments are supported by high performance in-built operating software for control and recording purposes. In most observatories around the world, data is being stored in its solid-state memories. These systems usually support on-demand data download onto PC hard-disk or removable media by the operator. Data is analyzed and maintained either at the observatory or transported to "local seismic station" through physical storage media such as Tape, floppies, hard-disk, etc. This has been the older conventional approach.

Each observatory or monitoring station is called a "node" in networking language. Usually, they are un-manned stations situated remotely at the locations of interest or activity.

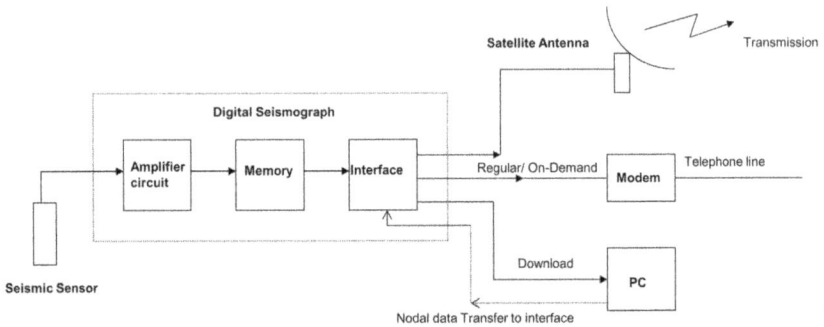

Fig [1]: Existing Seismic Data Acquisition System connected to a seismic station

The technological revolutions have converted these nodes into advanced seismic stations which are equipped with advanced PC controlled or PC interfaced digital seismographs capable of communicating with the "local seismic station" via modem or satellite. A typical setup is shown in Fig [1]. Depending upon the architecture of these seismographs, the data may be stored locally in its memories for subsequent transport physically or could be instantly transmitted online or could be available with on-demand offline transmission to a central data collection station.

This data is transmitted by instruments itself at suitable intervals under its own control or is transmitted on-demand upon receiving a request from "remote station." This transmission is generally done through local telemetry systems working on microwave frequencies or via dial-up connection through telephone lines. Local telemetry has been a popular option of transmission of data from the nodes to the central station. Satellites Telemetry systems have also come into use, which receives the data from the remote nodes via satellite link.

There have been many efforts and architects for networking of these remotely situated seismic nodes to make "local seismic station" (for larger areas) and networking of "local seismic station" to make a "regional seismic station". Data from these geographically distributed "nodes" can reach the "local seismic station" instantly through microwave/satellite links or can be on-demand downloaded through a modem. However, there is little or no data transmission or forwarding among various local or regional stations even if they are connected to a main "central seismic station." The current structure of a networked "nodes" across a region is shown in Fig [2].

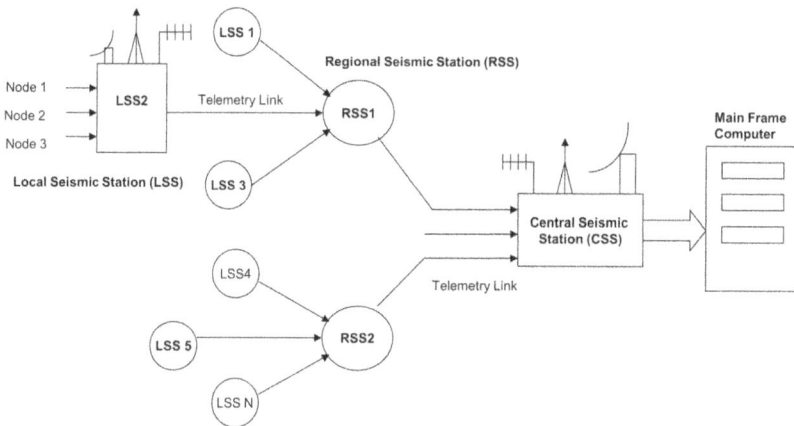

Fig [2]: Existing Telemetry Based Networked Seismic Monitoring Systems

With networking and telemetry technology advancement, Real-Time monitoring system (RTMS) has been implemented in many regions. In RTMS, data is continuously transmitted from the seismic nodes to a "central seismic station" via local/regional seismic stations. The online data from many closely situated and remotely situated nodes are available to experts and users at "Central Seismic station." With diverse data, better inferences can be drawn, and better analysis of seismic activity can be carried out. The remote seismic station works as a drop, where data is continuously coming in and getting out to a network of RTMSs to make a "national Seismic Network".

III. LIMITATION IN EXISTING NETWORKED SYSTEM

The existing instrumentation networks have been working upon a wide variety of architectures. The data transmission modes included microwave, UHF, dial-up, Satellite links depending upon the distances and applications. These systems equipped the experts with diverse data available from the distributed network on a common platform and hence better inferences could be drawn. Still, this data is not available instantly to the experts and users in the other part of the world. The major problem with these systems is that user worldwide could not access the data recorded by the observatories online or offline.

What could be accessed is the pure data and not the information contained in the data. Information is the most important part of this data; otherwise, data cannot be utilized in any practical manner. A major shortcoming has been that data not properly converted into information with the help of extensive analytical tools such as visualization, contouring, seismic interpretation, statistical calculations, and mapping, etc. Further, the database of the entire data is maintained in the conventional methods, which do not support query processing and search engine-based database management.

The extensive efforts in data collection go waste when data is not converted into information and information inferred so as is not presented to users the entire world in a universally accepted mode of presentation. The drawback of the conventional systems has been that a universal approach in providing the graphical user interface has not been tried on global scales. The existing systems have been catering to needs of the limited user segments. Communication and networking technology advances do have been integrated into the system. However, the user-friendly interface and easy but fast retrieval of information need elaborate attention.

The entire real-time monitoring networks have been effectively interfaced to communication links worldwide, and many agencies have also been set up who shares the data under some agreement. However, the present system also poses a problem of compatibility when the issue of globally interconnecting such remote stations is raised. The incompatibility of data format, non-availability of some common communication platform and intercontinental data transportability have been causing problems.

Even not all the seismographs support the data transmission and networking via telemetry or WAN. Data is maintained individually only and provided only when demanded by the central station. In such a situation, data from these nodes is dumped to physical media and transported back to the central station. This process may take quite some time and data, which may be very crucially needed, is not available online or instantly.

IV. INTEGRATION OF CURRENT SYSTEM TO INFORMATION TECHNOLOGY

The understanding of geological phenomena to make predictions of substantial economic and societal value remains an important goal for geo-scientists all over the world. Geoscientists need elaborated information to predict the phenomenon like an earthquake, snow-avalanche, snowfall, flood, and geophysical climate. These forecasts will be used in resource planning and avoiding harmful effects such as life and material loss caused by them.

Information technology has affected the way one gets the information. The internet is being taken as the fastest and universally accepted user-friendly mode of information retrieval from the wealth of data available at thousands of servers situated all over the world, connected over the net via LAN and WAN. Information technology has to be integrated with complex geo-scientific instrumentation for data mining, visualization, and analysis to form an information system namely Geo-Scientific Information System (GSIS). This system has to be highly reliable, and it should prompt information to the users and the experts all over the world through the fastest possible mode. Online & interactive availability of information would be the major advantage of this system, which could be achieved only with implementing proper internet tools and Information Technology.

The developed system provides a Knowledge-based search and analysis engine that allows users to obtain data whether they know exactly what to retrieve and let them identify, among available data, a significant correlation, trends worthy of further analysis and assess the data to be retrieved. This support allows additional communities such as process scientists and applications users to access the data.

The information technology implementation is innovative and scalable, substantially using latest widely accepted advancements, and relies on web technology with a multi-tiered client-server system architecture that enables easy access for the user. The user can access the desired information from Main Information Server through the web server using a standard internet browser and file transfer protocols.

V. CONCEPTUAL ARCHITECTURE OF GSIS SYSTEM

From the discussion elaborated above it is clear that the very first design step has to be the installation of modem/VSAT/Iridium Handset/Cellular interface card in the Digital seismographs at each observatory along with network connectivity. Unlike the previous case, we would like a powerful PC interface with Digital Seismograph with PC having all communication interfaces, as shown in Fig [3].

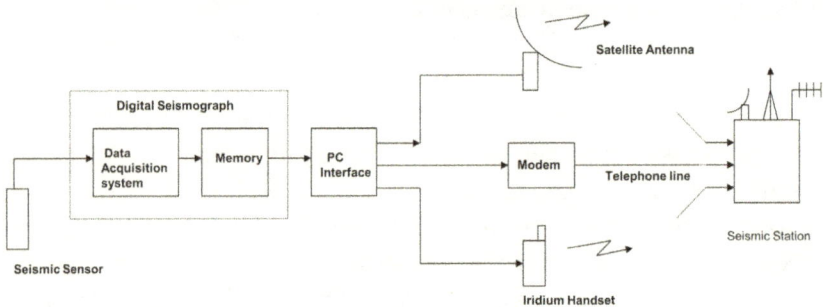

Fig [3]: Networked PC based architect at node

This will make seismograph interdependent of the communication and networking load. The networking capability can be integrated into the setup very easily using state-of-the-art PCs. In the earlier conventional case to PC was an inherent part of the node, but its use is limited to data transfer or downloads to transportable media only. However, in this case, PC will act as nodal database point, which will not only process the data, it will analyze using an analysis tool, perform the extensive calculation, computation and organizes the inferences as a unified database under the acceptable standards for such a network.

The connectivity of this database and PC can be provided by dial-up connection through telephone lines, Microwave Telemetry, Satellite or WAN links. This will enable to form a centrally controlled efficient network of instruments. The existing systems in which the interface card cannot be installed may be needed to reconfigure. As a design alternative, these instruments can be interfaced to a PC with communication facilities.

The underlying concept of the system is to establish distributed geo-seismic observatories, each acting as "node." Sometime due to technical problems, it may not be possible to make each measuring center as a node. So, several sub-nodes can be coupled together to make a node.

Each node in itself would be a PC based seismograph with standardizing data conversion, formatting, communication capabilities and to put on a network. Each node would be available with continuous acquisition of seismic data in some standard format, eg. SUD, ASCII, etc depending upon the instrument used. Each node will have software utility for converting these data formats into a mutually agreed single universally accepted data format and stored at a predefined location in this PC node.

There may be similar nodes in many numbers spread over the whole geographical area. The internal architect, design, and setup of each instrument may be different. The stress is on the same format of "data" across all nodes.

The instrumentation systems at various nodes are connected to a local or satellite seismic telemetry stations and continuously sending recorded data. Many such local telemetry stations may be connected to a regional station. The regional station can be further connected to a National Seismic Station, where data from all the seismic nodes is being dropped into.

The data received at the national seismic station is maintained in exhaustive Relation Database management systems on the main server. This RDBMS is created because of standardized analysis tools and software performing exhaustive correlation, 2D & 3D visualization, Mapping and contouring, Statistical calculations, seismic waveform visualization & interpretation, high-performance numeric computation, etc.

Now where to store the data? Each of the nodes has a data storage capability and data can be stored on each of the local seismic stations too. A secondary server at each of the local seismic station would store the data from various nodes in a unified manner.

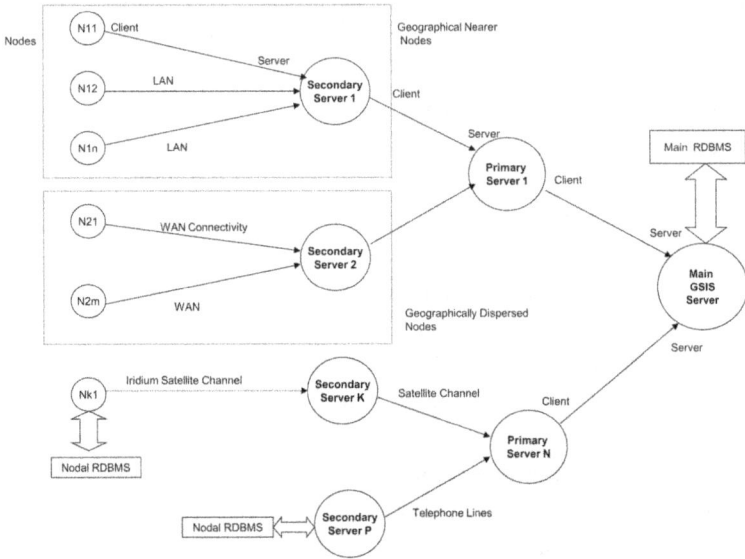

Fig [4]: Networking of Servers in Server-Client Peer to Peer Architecture

Here a multi-tier server-client architect would be best to be implemented. Many nodes can be coupled to make a secondary server. To this server, all the other connected nodes shall work as clients. Further many secondary servers shall be connected to make a primary server. To this primary server, all the secondary servers shall act like clients while at the same time these are acting like the server to other nodes. This approach will be taken, as it is impossible to interconnect all distributed nodes to one main server directly.

Further, traffic load consideration makes it necessary to use a multi-tiered server-client system approach. Finally, all primary servers are connected to one main server. This main server will be accessible across the globe through the front-end web interface.

This approach is shown in Fig [4] which is actually a multi-tier client-server technology.

The main server would basically be an RDBMS server running high-end database like Oracle. Again, there are multiple architects available to implement the main server. The functions can be divided among the file handling server, web server, and a database server. The main server will be 24-hrs connected to an internet gateway.

Multi-dimensional query Processor (MDQP), an integrated front-end and backend database software interface of RDMBS would be needed to handle specific queries and information request by the user.

The user accesses the information through WWW browser on their PCs. The user may be needed to get authorization depending upon the user segment and type of information sought. Dynamic links in the web page provide the GUI interfaced content-based information presentation to the user. The user can select the desired information listed on the web page and click it to get expanded information. Search Engine based and specific query-based information is also provided to the user through the internet. MDQP will handle the processing of such queries. Online data, events information, and history database are also available to the user. Internet connectivity is available to the user through ISP or dedicated high-speed digital lines. Once the desired information has been obtained, the user can download either through WWW browser or through FTP tools.

VI. GSIS: KEY DRIVING FACTORS FOR SUCCESSFUL IMPLEMENTATION

While highlighting the problems in the existing system, user requirement is the primary driving force behind the design of any new system. Over the time the requirements and the needs of the users have changed, and this system is expected to fulfill their needs.

a) *Presentation of information inherent in the data and analysis as*pect is one of the major requirements of the users. It is possible because of the close integration of data and services with analysis software tools. Some new data analysis and multi-user software packages may be needed to develop in the process of implementation information technology in the field of instrumentation. It has to assist users further to produce enhanced information and reports by integrating complementary data sets in a centralized point of access. A set of standardized and universally accepted analysis tools has to be applied to the diverse data sets at the server to create ease of use and compatible data exchange. The graphical plots and tables along with the correlation coefficients, means, standard derivations, and other statistical parameters derived from the content-based browsing may form a resulting report.

b) The *use of available Web browsers* is also a major requirement of the proposed system. Thus, the interface between users and this distributed system has to be via the Internet. The data sets may be downloaded/transmitted via FTP or WWW over the network. Concerning the communication protocols among sites, survey and proper study are needed to carry on selecting proper protocol standard and to make a more open architecture to systems outside to create more opportunity for inter-operation.

c) *Implementation of the Query Engine* is fundamentally driven by Variability of geological parameters in this approach. The geo-scientific system science aspects lead to several user queries and access scenarios related to observational and model output data sets. The overall system is to be designed to enhance the current ongoing capabilities to serve the specific geoscience applications and provide an innovative query engine to the users.

d) The *flexibility of the system architecture at the user end* is of course expected, which should allow organizations and users at their terminal end to choose among various standard network, hardware, and operating system configurations that range from small, inexpensive system to relatively expensive workstations. The extensibility of the architecture should allow the users to upgrade their systems as their needs and system capabilities evolve needed to meet specific research and educational requirements. To provide this flexibility, the main seismic station server has to be compatible to communicate with various types of computers and operating systems viz. IBM PC on MSDOS operating system and MAC personal computers; Windows 9X/XP operating system terminals, DEC VAX stations that use the

VMS operating system; and Sun Microsystems workstations working on UNIX operating system etc.

VII. WORKING OF GSIS

A. *Communication Between Nodes & Servers*

Each of the node, the secondary server or primary server can act as a client and the server. This is a peer-processing mode of data exchange. Anticipating the huge volume of seismic data generated, it may not be possible to handle and process the data at one main server only. So, a trade-off has to discovered to process the seismic data at each nod to reduce the traffic load. Conversion of recorded seismic data into the desired format is to be done at the node(s) itself. The platform of the node may be any of PC-AT, PS/2, DEC VAX or SUN working on any standard operating system, but it has to be made compatible with the data communication process involved using universally accepted protocols.

The PCs at different nodes are configured as a dial-up server and/or data transmitting node. The connection from nodes to the main server, through secondary and primary servers, is accomplished by a LAN/WAN. LAN connectivity is to be used only when some nodes are in proximity to each other, and a single administrating agency is managing these multiple nodes. Main stress in this system is placed on exploiting existing WAN connectivity with enhanced featured for this specific application. The WAN will operate via satellite, microwave link or dial-up connection with standard communication software and protocols.

At some nodes, where general communication facilities are not available, data retrieval via iridium handset phone is also considered. The seismic data and results available at various nodes are transmitted to Main Server routed through the tiered configuration of nodal database servers.

Selecting the right media (through which signals shall travel from nodes to the server and then to users such as Direct Cable, Telephone, Microwave, Satellite, Cellular for the network,) for the network is a crucial and important task in this system. No matter how powerful the servers are and how high individual capacities of the nodes are if a fast and efficient communication media do not connect them, then the network achieved from that place can simply be a waste of time, money and effort. The primary criterion for evaluation of transmission media of this system is the application intended to use on the network. Other important parameters are bandwidth as some nodes on the network might need larger Bandwidths (say 10s of Mbps) for example downloading of images. Geographical Distance between the seismic nodes also decides the selection of transmission media, as each media has certain distance restrictions up to which it can carry a signal without attenuation. Easiness to install the cabling and associated equipment is another factor for consideration. Security issue may be very critical, where data transmitted is of strategic importance. Cost of development, implementation, and installation undoubtedly is the major factor that decides the type of media used in a network.

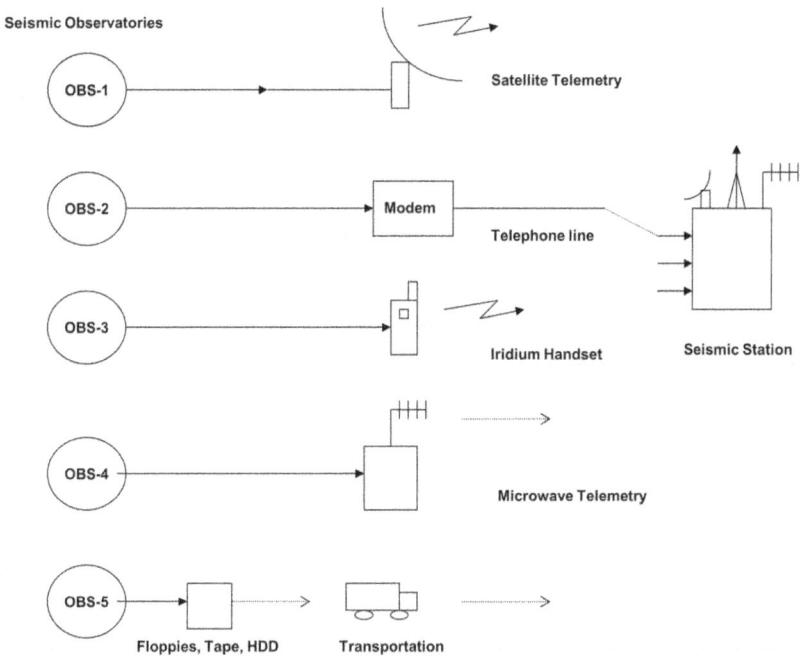

Fig [5]: Various Communication Links for Nodes to Seismic Station Connection

The aim is definitely to develop a low-cost GSIS system, so it could prove economical worldwide. Satellite media is proving cheaper these days for such applications. Provision of a telephone line communication is also an attractive alternative when an existing telephone network is to be exploited. Various communication media options are currently available for a node to a seismic station remote connection has been depicted in Fig [5].

B. Seismic Database Generation

At each of the node or secondary server, seismic data is collected and is properly formatted for storage. The inferences and information extracted from this data can also be stored on each node or secondary server with properly classified file names.

The option of having nodal databases or centralized database is governed by many other factors like topology of the network, individual network leg speed and processing capabilities available with each node as well the type of application for which data is being stored. Sometimes, the main Server may contain only highly important and crucial processed information frequently accessed all over the world. In such cases, most of the raw data & related information may reside in nodal-databases with the local or regional telemetry stations and linked to the main server database.

The nodal database may exist with individual "nodes," whereas "local seismic stations" may have secondary servers, "regional seismic stations" having the primary server and at national level or continent level main server can be set up. So the discussion here is regarding the network of servers under server-client topology, which equivalently means a network of seismic stations locally or regionally. The typical server-client concept was depicted in Fig [4]. However, perhaps this networking provides database creation and management facility at various levels also, instead of just pure data storage as was the case with conventional telemetry based networked nodes and seismic stations.

If the volume of data from each node is too large to be handled at main server, it may be converted into the nodal-relational database with exhaustive correlation techniques, visualization techniques, mapping and contouring, interpretation and statistical calculations and modeling at nodes itself. The exhaustive information to be stored along with data may include the seismic waveform analysis/visualization and images.

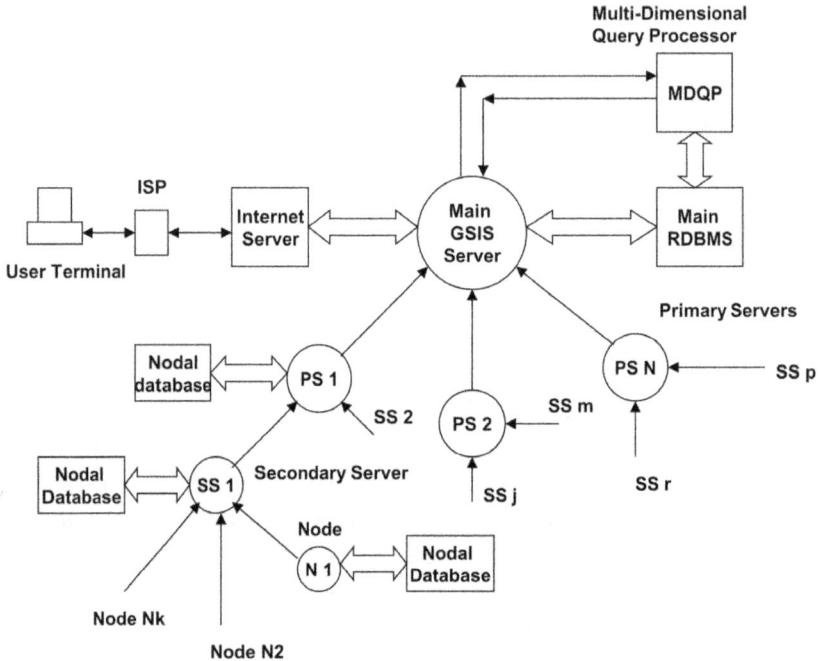

Fig [6]: Seismic Database Generation Through a network of Servers and databases

These distributed nodal-databases are interconnected together using web technology at the Main server, through client-server architecture, as shown in Fig [6]. The dynamic address link to proper information is defined at the server for prompt access to information.

A centralized Relational database containing exhaustive geo-seismic and the geophysical information is the core of a GSIS system design. The Relational database is generated because of various statistical, analysis-syntheses and modeling techniques but, such information is stored after correlating the individual node's data. The database at the Main server contains both raw and interpreted information for all seismic events, event histories and reports, online data, seismic profile studies of the area and a large number of seismic surveys carried out by the nodes.

Relational data model needs to be designed on standard guidelines using scalable highly secure and integrated database platform such as Oracle. This provides solutions and support to integrated multi-disciplinary projects in other fields as well and allows a standardized data transfer between the database and other applications helped by many popular software

RDBMS needs to be maintained over a high-performance workstation (Main Database server). This database also integrated to a multi-dimensional query processor (MDQP), which will process the specific queries by the users related to data or information contained in the database. RDBMS & MDQP are accessible via web page lying at any internet or web server.

MDQP will handle the specific queries which user may put up. It will supply the content-based criteria on a parameter, direct the Main Server to explore its database for the requisite information. The queries are pre-processed by Multi-Dimensional Query Processor that accesses the specific parameter, statistical summary, data, and predefined tables. A query may be multi-dimensional in nature; i.e., it specifies the ranges of several spatial and temporal dimensions and requires the application of analytical tools on the selected data associated with the chosen ranges. This allows for a quick look at the data and return of results to the user. The key to the scalability of this system lies in the fact that data sets need not be transferred among different nodes for an interactive query

C. User Access to GSIS Over Internet

The user needs to be connected to the mail or internet server through any ISP (Internet Service Provider), which would, in turn, configure the gateway for the user to the outside world through an access device. The user has to install software and communication hardware to facilitate connection and to access various internet services at its own terminal running any operating system. Individual user, department or agencies subscribing the seismic data may manage internet access via dedicated/leased line connections or higher speed digital transmission lines.

The communication interface from user to Main Database Server is provided through standard internet protocols. SLIP (Serial Line Internet Protocol) and PPP (Point-to-Point Protocol) both allow the user to dial-up and actually run internet applications on his own computer via a regular phone line and a modem or satellite/cellular etc. These high-end connections call for a phone line (or satellite connectivity), a fast modem, and a computer capable of running TCP/IP and Internet Access. TCP/IP (Transmission Control Protocol/Internet Protocol) software makes every system on the internet interoperable and makes it capable of carrying high bandwidth, interactive multimedia internet applications, e.g., graphics, sound, animation. A typical setup of a user with the main server through the internet is shown in Fig [7].

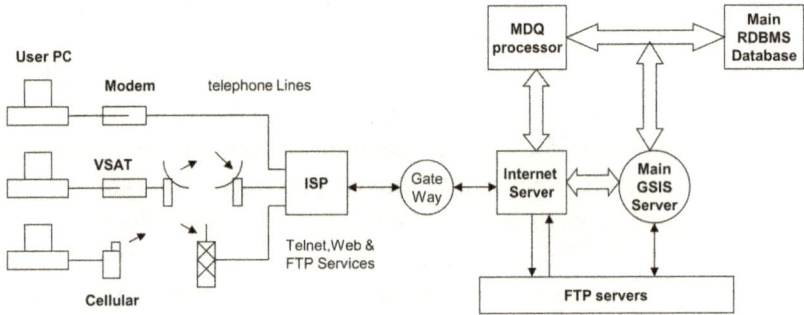

Fig [7]: User Access to the main database through the Internet

Other than normal operational software, the user machine would be requiring no special software for data analysis as the information user would get from the server via the web interface would be formatted reports, analysis results, and tables. The actual analysis will be done by the MDQP at server depending upon the query. However, the advanced user may have sophisticated analysis tools to further make data analysis depending upon his desired application.

D. Web-Based GSIS User Interface

The user interface is provided by Worldwide Web Browser which gives an easy-to-navigate interactive graphical interface for searching, finding, viewing and managing information & documents over a network. This global medium is gaining popular acceptance faster than any other communication medium in history. It is no doubt the primary aim of this system to exploit the current popularity and advantages of the web in integrating information technology with the geo-seismic instrumentation system.

The main idea in information browsing in this system is to generate a dynamic web page over Worldwide Web. The dynamic linked web page is updated continuously as per the continuous information arriving from every node. The parameter statistics keep on updating in such a web page when especially the incoming data is coming in real-time mode and is needed to be accessed in real-time mode. It helps in instant access of current data at a different node of interest and flashes the instant analysis of data by the experts.

To access information, a user can access the main web page of the GSIS system through internet. This main web page of the system is written and formatted in HTML (Hyper-Text Markup Language), and words are linked to connect different sites and pieces of information to one another. Links embedded in words or phrases allows the user to select relevant text and immediately get related information and multimedia material. The information may be presented using a variety of media such as text, graphics, audio, video, animation, image or executable documentation.

E. Modes of Information Browsing at GSIS

The process of information browsing access of the main web page of the system which prompts the user with the following three modes of information presentation.

First is *Content-based Information* as provided by the web page. The web page of GSIS presents the user with a list of hyperlinked contents & words. The hyper-linked datasets include listing the location of nodes, Date wise event records, Epicenters & Timing of seismic events, a major event listing, available waveforms & digital data files, Area Coverage, maps, survey reports, seismic news updates, and online data access options. The hypertexted index is linked to the properly addressed information page, which may be lying with the same or other internet server or the main server of the system. After selection of information content, the connection is established to the pointed address. The dynamic web page accesses the information from a designated server and presents it to the user.

Second is *Search Engine based Information access* in which specific keywords search will be carried out in the entire seismic database and nodal databases. In this interface, keyword items are interconnected and form a link to dynamically generated a web page. A user is presented with a list of the phenomena and relevant specific parameters linked to each phenomenon. Just from one keyword, the entire database is searched for linked data, and relevant information is invoked and presented to the user. Keyword items and their relationships are stored in a relational database with entities being Phenomenon, Parameters at the first level, followed by Phenomenon Instance, Specific Parameters, Instrument at the second level, Predefined Region, at the third level, and, finally, Statistical Summary, Data Format, and Data File, at the lowest level.

Query-based Information Access is the third mode of information presentation. The user may put up specific queries, which supplies content-based criteria on a parameter, direct the Main Server to explore its database for the requisite information. The queries are pre-processed by Multi-Dimensional Query Processor that accesses the specific parameter, statistical summary, data, and predefined tables. A query may be multi-dimensional in nature; i.e., it specifies the ranges of several spatial and temporal dimensions and requires the application of analytical tools on the selected data associated with the chosen ranges. This allows for a quick look at the data and return of results to the user. A user can choose at first instance, the day of interest and asks what the seismic events that occurred on that day. The system returns dataset dates satisfying above conditions. The user can further refine his search criteria by mentioning only events crossing 4.5 Ritchet Scale at the specified date.

If the presented information suits his requirements, he can download or ask for spectral analysis of the event or can order complete data sets for the above date.

F. Information Download from GSIS

Internet tools like Telnet and FTP are available to users for accessing data when the location of data is known. Telnet program offers a way to log into the server and work from another computer. By logging into another system, users can access services/data or files thereon with authorization restrictions. FTP, or File Transfer Protocol, lets users download files from another computer system on the internet, or on a local network. It can also work the other way around with users/nodes also transferring their own files to other computer systems. The user can only access computers set up as FTP servers. A list of relevant FTP server containing seismic information has to be maintained.

VIII. CONCLUSION

Even within country non-uniformity in geological manifestation has been observed. No doubt distributed measurement network is the only solution left for such non-uniform parameter measurement. The overall tasks include the implementation of low-cost information technology for faster access to remote information in the centralized mode and distributed data access. The vision can serve as a model for a larger overall data information system associated with future Geological Monitoring datasets and their distribution. The reader can go through the reference and will find that GSIS systems have been employed in many parts of the word successfully for a unified database creation.

The system here has been described referring to seismic applications. Similarly, it is extended to Snow Hydrological Studies in deep snowbound mountain areas and carrying out related forecast and analysis. The design approach can serve as a model for a larger overall data information system associated with future Geological Monitoring datasets and their distribution.

The system design of GSIS involves the integration of information across many domains such as instrumentation systems, Networking, Relational Database management, and information technology in a distributed fashion. Emphasis is given on the ability to move rare data (recorded at geo-scientific observatories) effectively between widely dispersed computer systems over the net using standard protocols and data models. The processed data is presented to users in interactive and user-friendly mode. Diverse Computer software/platforms like RDBMS, Geographical information systems(GIS), 2D/3D Interpretation &

visualization tools will be used for the geological parameters mapping of earth's manifestations viz. seismic events, hydrological phenomenon etc.

A consortium of experts from all over the world has to be established to lookup the matters related to standardized, future expansion, security and strategic use of data once the GSIS system comes in the picture. They can further act as solution providers for the design of the overall system taking into view the aspects of upgradability and technology revolution.

IX. REFERENCES

[1]. L.A. Trenishand M.I. Gough, 1987: *A software package for the data independent management of multi-dimensional data.* EOS transactions, Am. Geophysics U., 68, 633-635.

[2]. David W. Fulker, *Seminal Software to Analyse and manage Geo-Scientific Information,* Unidata Program Centre.

[3]. Menas Kafatos, Ruixin Yang, *X.* Sean Wang, Zuotao Li and Dan Ziskin, *Information Technology Implementation for a Distributed Data System Serving Earth Scientists: Seasonal to Interannual ESIP.*

[4]. M. Kafatos, Z. Li, R. Yang, et. al., 1997, "*The Virtual Domain Application Data Centre: Serving Interdisciplinary Earth Scientists,*" Proceedings of the Ninth International Scientific and Statistical Database, 264-276. IEEE.

[5]. K. Kovar & H. P. Nachtnebel, 1993, *Application of Geographic Information Systems in Hydrology and Water Resources Management,* Proceedings of the HydroGIS 93, Vienna.

[6]. E. Mesrobian, R. Muntz, E. Shek, S. Nittel, M. LaRouche, and M. Kriguer, 1996, *OASIS: An Open Architecture Scientific Information System,* Proc of 6th International Workshop on Research Issues in Data Engineering (RIDE '96) p. 107, New Orleans, IEEE Press.

[7]. E.C. Shek, E. Mesrobian, and R.R. Muntz, 1996, "*On Heterogeneous Distributed Geo-scientific Query Processing,*" Proc. Sixth Int'l Workshop Research Issues in Data Engineering, IEEE CS Press, 1996, pp. 98-106.

[8]. K. Beuhler, and L. McKee (eds), 1996, *The OpenGIS Guide: Introduction to Interoperable Geoprocessing,* Open GIS Consortium Inc., Wayland, Mass., 1996.

[9]. G. Graefe, 1994, "*Volcano: An Extensible and Parallel Query Evaluation System*" IEEE Trans. Knowledge and Data Engineering, Vol. 6, No. 1, pp. 120-135.

[10]. S. Fushimi, M. Kitsuregawa, and H. Tanaka, 1986, "*An overview of the system software of a parallel relational database machine GRACE,*" in Proc. Conf. Very Large Data Bases (Kyoto, Japan), pp. 209-219.

[11]. T. Keller and G. Graefe, 1989, "*The one-to-one match operator of the Volcano query processing system,*" Oregon Graduate Centre, Computer Science Tech. Rep., Beaverton, OR.

[12]. Bath, A. 2005, *Nationwide collection, recording, provision of Geo-Scientific data*, Beak GmBH, URL: http://www.beak.de/pdf/news/seee_presentation_2005.pdf.

[13]. Erik World, 2005, *dGB-GDI system: Software for analysing and quantifying geo-scientific information*, URL: http://www.sintef.no.

[14]. Robert C. Yoder, 1996, *Design Considerations for Implementing Complex Client/Server Applications: A Case Study of Octree-based 3-D GIS*, University of Albany, URL: http://www.albany.edu/~sysrcy/dissertation/PR-11.TXT.

[15]. P.N. Moir, M.A. Williamson and S. King, *Integrating geo-scientific data, knowledge and interpretation for the east coast basins of Canada*, Geological Survey of Canada Atlantic, URL: http://agcwww.bio.ns.ca/hcmp/newsltr7/integrat.html.

[16]. Paul D. Amer, 1982, *A measurement Centre for NBS Local Area Computer Network*, IEEE Transactions on Computers, Volume C-31, Number 8, pp. 723-729.

[17]. NASA, 1997, *"NASA Selects Earth Science Information Partners"*, NASA Press release. *URL:* http://www.nasa..gov/releases/1997/.

[18]. Doty, B. E., Kinter III, J. L., Fiorino, M, Hooper, D., Budich, R., Winger, K., Schulzweide, U., Calori, L., Holt, T., and Meier, K., 1997, *"The Grid Analysis and Display System (GrADS): An update for 1997,"* 13th Conference on Interactive Information and Processing Systems for Meteorology, Oceanography, and Hydrology (American Meteorological Society, Boston, pp 356-358.

[19]. Date C.J., 1985, *An Introduction to database management System*, Vol 1, Narosa Publishing House, New Delhi.

[20]. R. Löwner, A. Souhel, 2004, *The Sahel-Doukkala Information Network (SaDIN): A regional online geo-information system*, URL: http://www.enviroinfo2004.org/cdrom/Datas/loewner.htm.

SOFTWARE TOOL FOR SEISMIC DATA RECORDER AND ANALYSER

Kumar, S, Attri, RK, Sharma, BK & Shamshi, MA, 2000, **'Software Tool for Seismic Data Recorder and Analyser'**, IETE Journal of Education, vol 41, No. 1-2, pp. 23-30, https://doi.org/10.1080/09747338.2000.11415718

INDEX

ABOUT THE AUTHOR

Raman K Attri is a corporate business researcher, learning strategist, and management consultant. Masters in electronics engineering, he served as an electronics design scientist at a premier research organization. He has served at technical and product development roles at leading international corporations. As an engineer, he specializes in systems engineering of complex equipment, scientific instrumentation sensors and system design. His international professional career spanned over 25 years across a range of disciplines such as scientific research, systems engineering, management consulting, training operations, and learning design. With his technical and training background, he focuses on the competitive strategies to develop the technical workforce with higher-order troubleshooting and problem-solving skills at a much faster rate. He provides strategic consulting to the organizations by accelerating time-to-proficiency of employees through well-researched models. He holds a doctorate in business from Southern Cross University, Australia.

Speed To Proficiency
RESEARCH

Accelerated Performance for Accelerated Times

Highly-specialized know-how, learning, and resources to solve challenges of 'time' and 'speed' in performance at organizational, professional and personal levels.

Visit us at https://www.speedtoproficiency.com/

S2Pro© Speed To Proficiency Research is a corporate research and consulting forum that provides authentic guidelines to business practitioners to accelerate proficiency of their workforce, teams, and professionals at the 'speed of business'. S2Pro© publishes reports, ebooks, and articles exclusively related to accelerated performance, accelerated proficiency and accelerated expertise in individual and organizational context. Our extensive knowledge base of "how to methods" is derived from experience-based and practice-based observations, analysis/synthesis of existing research, or based on planned/focused research studies through a network of researchers who exclusively focus on 'time' and 'speed' metrics in the business context.

Speed To Proficiency Research: S2Pro©
A research and consulting forum
Singapore 560463

Website: https://www.speedtoproficiency.com
e-mail: rkattri@speedtoproficiency.com
Facebook: https://www.facebook.com/speedtoproficiency/
LinkedIn: https://www.linkedin.com/company/speedtoproficiency/
Twitter: https://www.twitter.com/speed2expertise
Google+: https://plus.google.com/101561704929830160312

www.ingramcontent.com/pod-product-compliance
Lightning Source LLC
Chambersburg PA
CBHW031905200326
41597CB00012B/537